Valuation of Hotels for Investors

David Harper

2008

A division of Reed Business Information

Estates Gazette
1 Procter Street, London WC1V 6EU

©David Harper, 2008

ISBN 978-0-7282-0522-2

Apart from any fair dealing for the purposes of research or private study, or criticism or review, as permitted under the UK Copyright Designs and Patents Act 1988, this publication may not be reproduced, stored, or transmitted, in any form or by any means, without the prior written permission of the Publishers, or in the case of reprographic reproduction, only in accordance with the terms of the licences issued by the Copyright Licensing Agency in the UK, or in accordance with the terms of licences issued by the appropriate Reproduction Rights Organisation outside the UK. Enquiries concerning reproduction outside the terms stated here should be sent to the Publishers.

The Publishers make no representation, express or implied, with regard to the accuracy of the information contained in this publication and cannot accept any legal responsibility or liability for any errors or omissions.

The material contained in this publication constitutes general guidelines only and does not represent to be advice on any particular matter. No reader or purchaser should act on the basis of material contained in this publication without first taking professional advice appropriate to their particular circumstances. The Publishers expressly disclaim any liability to any person who acts in reliance on the contents of this publication. Readers of this publication should be aware that only Acts of Parliament and Statutory Instruments have the force of law and that only courts can authoritatively interpret the law.

Copying in any form (including all hard copy and electronic formats) is strictly forbidden without written permission from the Publishers, Estates Gazette, a division of Reed Business Information.

Cover design by Rebecca Caro. Image supplied by Rex Features.
Typeset in Palatino 10/12 by Amy Boyle
Printed by Progress Press, Malta

Contents

Acknowledgements vii
Dedication. viii
Table of Cases ix

1 Things to Consider When Buying a Hotel and How to Avoid Potential Pitfalls 1
   Introduction. 1
   Initial considerations when first thinking about buying a hotel 1
   The various types of hotels that can be purchased. 4
   Grading of hotels 10
   Finding the right property 10
   Affordability 11
   Finance. 12
   Reviewing the existing business 12
   The importance of due diligence 12
   Summary 13

2 Understanding the Trading Potential of the Hotel. 15
   Introduction. 15
   The inspection. 16
   The local hotel market. 22
   Analysis of the accounts 27
   The interview 37
   Brand names 41
   Independent analysis of the information 41
   Summary 41

3 Other Enquiries Required to Assess the Value of the Hotel 43
   Introduction. 43
   Title and tenure. 43
   Statutory enquiries. 45
   Statutory enforcers — operating licences. 46
   Summary 49

4	**Fundamental Principles Behind the Hotel Investment Market**	51
	Introduction	51
	What is an investment?	51
	Who are the buyers and why are they buying?	52
	What types of hotel investments are available?	53
	Valuation methodology	54
	Summary	55
5	**The Purpose of the Valuation and Statutory Guidance**	57
	Introduction	57
	Introduction to the *Red Book*	58
	Purposes for valuation	63
	Disposal/purchase — transaction advice on capital values	64
	Capital value — secured lending purposes	67
	Capital value — company accounts purposes	72
	Capital value and calculations of worth — internal company purposes/strategic advice	77
	Rental levels — transactional advice	77
	Annual valuation — for taxation purposes	78
	Valuations for depreciation in company accounts	81
	Summary	85
6	**Methodology for Capital Valuations**	87
	Introduction	87
	1. The comparison method	87
	2. The investment method	88
	3. The profits method	88
	4. The residual method	88
	5. The contractors method	89
	Hotel valuation methodology	91
	The profits method (and the investment method)	93
	The comparison method — comparable evidence and prices per bedroom	106
	Check methods, and determining factors in the purchase of hotels	115
	Valuations in exceptional circumstances	118
	Summary	119
7	**Yield Selection and Valuation Multipliers**	121
	Introduction	121
	Vacant possession properties	122
	Yield selection for investments	124
	The key terms for each type of investment	129
	Other methodologies for assessing the appropriate yield	134
	Problems with comparable evidence	136
	Summary	138
8	**The Treatment of Capital Expenditure**	139
	What is capital expenditure?	139

	Timing of capital expenditure	145
	Summary	149

9 Methodology for Assessing Market Rent ... 151
Introduction ... 151
Comparable method ... 154
Profits method ... 158
Rental levels as a percentage of total investment ... 163
Rents per meter or square foot ... 164
Summary ... 164

10 Methodology for Calculating Rateable Values in the UK ... 165
Introduction ... 165
What are rates? ... 165
Assessing the taxation liability of a hotel ... 165
Transitional relief ... 166
Method of assessing rateable value ... 167
Summary ... 170

11 Site Values and How They are Determined ... 171
Introduction ... 171
Basic methodology ... 171
Summary ... 178

12 Valuation with Special Assumptions for Secured Lending Purposes ... 179
Introduction ... 179
Summary ... 183

13 IFRS 15 Apportionments ... 185
Introduction ... 185
What is depreciation? ... 185
Summary ... 188

14 Concluding Summary ... 189

Glossary ... 191
Index ... 195

Acknowledgements

There are many people who were kind enough to provide me with help in writing this book from within the hotel sector and, indeed, from other property sectors. Everyone's contribution was important and, as such, my thanks are in alphabetically order.

My thanks must go to: Richard Abbey; Jamal Abraham; Neil Adcock; Guy Addison; Troy Anson; Adrian Archer; Lesley Ashplant; Derek Baird; Nigel Barrington; Paul Barton; David Battiscombe; Nick Boyd; Daniel Braham; Louisa Browne; Iain Bryson; Richard Bursby; Nigel Buxton; Richard Candy; Robert Chess; Jane Christensen; Sandra Clark; Stewart Coggans; Alan Cole; Jon Colley; Graham Craggs; Chris Dallison; Alan Dorrington; James Effingham; Mark Ford; Gill Fuller; Chris George; Matthew Grodd; Paul Harrington; Pippa Harrison; Tom Hawley; Erlend Heiberg; Tim Helliwell; Andrew Horder; Mark Horne; Anna Hwang; Brian James; Keith James; Philip Johnston; Rupal Katechia; Catherine Kelly; Stephen Laud; Simon Lewsey; Ben Martin; Paul McCartney; Sam McClary; Alister McCutchion; Nick Newell; David Ramsell; Carl Ridgeley; Rupal Salinas; Shripal Shah; Bob Silk; Paul Slattery; Bill Smillie; Serge Sonti; Tim Stoyle; Rod Taylor; Steve Timbs; Julian Troup; Justin Webb; Mark Whitfield; Steve Williamson; Katie Wright and Huw Zachariah.

Special thanks must go to Alistair Brooks for his help with certain sections and with the photography, Catherine Harper for hours of typing, and Lorelle Anson for a constant supply of tea.

I also thank Alison Bird, Paul Sayers, Isabel Micallef, Amy Boyle and EG Books for their help and patience.

David Harper

For Matthew and Nicola

Table of Cases

Banque Bruxelles Lambert SA *v* Eagle Star Insurance Co Ltd [1995] QB 375; [1995] 2 WLR 607; [1995] 2 All ER 769; [1995] 1 EGLR 129; [1995] 12 EG 144; (1995) 69 P&CR D37

Corisand Investments Ltd *v* Druce & Co [1978] 2 EGLR 86; (1978) 248 EG 315

Harmsworth Pension Funds Trustees Ltd *v* Charringtons Industrial Holdings Ltd (1985) 49 P&CR 297; [1985] 1 EGLR 97; (1985) 274 EG 588

London Borough of Camden *v* Civil Aviation Authority [1980] 1 EGLR 113; (1980) 257 EG 273

Merivale Moore plc *v* Strutt & Parker [1999] Lloyd's Rep 734; [1999] 2 EGLR 171

Pivot Properties Ltd *v* Secretary of State for the Environment (1980) 256 EG 1176; (1981) 41 P&CR 248

Plinth Property Investments Ltd *v* Mott, Hay and Anderson (1979) 38 P&CR 361; [1979] 1 EGLR 17; (1978) 249 EG 1167

Ravenseft Properties Ltd v Davstone (Holdings) Ltd [1980] QB 12; [1979] 2 WLR 898; [1979] 1 All ER 929; (1979) 37 P&CR 502; [1979] 1 EGLR 54; (1978) 249 EG 51

Singer & Friedlander Ltd *v* John D Wood & Co [1977] 2 EGLR 84; (1977) 243 EG 212

Zubaida (t/a Elley's Enterprises) *v* Hargreaves [1995] 1 EGLR 127; [1995] 09 EG 320

Things to Consider When Buying a Hotel and How to Avoid Potential Pitfalls

Introduction

In simplistic terms, the value of a hotel is the price that a prospective purchaser will pay for the property if it has been effectively marketed. There are many things that determine the price that potential purchasers will pay for the property, including potential earnings from the property, the need to provide geographic coverage to maintain the integrity of a brand and potential increases in value of the asset over a period of time. A big proportion of larger hotels are purchased (and sold) with the above criteria in mind, where returns on capital employed are considered to be just as important as strategic ones.

However, the majority of hotels that are sold in Europe are smaller properties and these are not necessarily purchased for the financial returns they can generate. This chapter is not aimed at chain hoteliers or investors who will have their own specific criteria that will lead to their choice of property; instead, it is predominantly aimed at less experienced hotel purchasers who are looking to buy smaller hotels, including those considering purchasing a property as a "lifestyle purchase" (those who are looking to swap the hurly burly of urban life for the perceived "gentler paced" life of a country house hotelier).

This chapter outlines the main considerations that the purchaser should take into account when buying a hotel, types of ownership, how to find the right property and outlines the importance of professional help throughout the process, and the types of due diligence that could be needed.

It is important that the valuer understands these considerations fully and is able to take them into account. Properties that meet the needs of a larger number of potential buyers could well generate more competition leading to higher purchase prices.

Initial considerations when first thinking about buying a hotel

There are many questions that a potential hotel purchaser needs to ask themselves before starting to look into buying a hotel and it will be a useful exercise for both the potential purchaser and the valuer to go through some of these questions here.

What does the buyer want from the purchase?

This must be the most important question — what exactly is the buyer after? If the answer to this is unknown then it is highly unlikely that the buyer will be looking in the right places or at the right properties to find the perfect property to meet their requirements.

Some purchasers may want to escape the pressures of city life, some may want to move to a different location, others may want to earn a substantial income, some will be buying the hotel as a stepping stone to a better property, others may be after a "tax-free" lifestyle, others may be looking at the business as a pension fund in a few years' time, whilst yet others may be looking to start a larger hotel company.

It is probably a useful exercise for a potential purchaser to write a list of all the things that are important to them, those that are less important and those that do not matter at all. This can form the basis for drawing up a shortlist that can help you determine the type of property that will be most suitable.

Is there a specific location?

As with all property, but especially for hotels, the most important factor in determining value is location. This is also true of "lifestyle" hotels — the most desirable locations from the potential purchaser's perspective will generally be the most expensive.

The amount of trade that can be generated in a particular location, combined with the desirability of the location from an aesthetic viewpoint will ensure that any properties benefiting positively from both will be well sought after. Quieter, less sought after areas may still provide less expensive options that could still meet the majority of the buyer's requirements.

It is then important to consider which location will suit the buyer's personal requirements, ie, is it close to your friends and family or, if they have a family with small children, does the location have suitable schools.

What type of lifestyle is the buyer after?

The reasons for purchasing the property is likely to affect the type of property the buyer is looking for. If they are after a relaxing lifestyle then a smaller property in a quieter location is likely to suit them better, whereas if the goal is to earn a substantial income then they would be better off looking at a larger property in a busier area.

Other key questions may include whether they want their private accommodation to form part of the hotel, or whether separate accommodation is required, and how large that accommodation needs to be.

What support and help will the purchaser have?

Running a hotel is not as simple as some people think, as John Cleese managed to show in *Fawlty Towers*. Without a support network in place, it can be very difficult to run a hotel entirely on your own.

A number of people planning to move into the small hotel market may feel they can run the hotel effectively on their own, but without adequate support it will soon become a chore rather than a pleasure. However the prospective purchaser must consider whether the property will be able to generate enough profit to pay for the support staff that the purchaser requires?

What are the buyer's strengths and weaknesses?

This is one of the key tests that will influence how the buyer should consider structuring the operations of the business. What are their particular strengths; are they good with customers and staff; are they better at marketing; are they adept at the finances and the accounts? Equally important, since all facets of the business are required to work well if it is to be a success is where their weaknesses lie, and how can these weaknesses be addressed?

Once the purchaser has considered what it is they want, where they want to be and what they will need to do in terms of improving their skills (or who they need to employ), they will be well placed to look at exactly what they can afford to buy, and how to raise the finance.

Types of ownership

Another fundamental factor that can influence the price of a property is whether it is freehold or leasehold.

In simplistic terms, a freehold property means that the interest in the land is held in perpetuity at no rental cost to the owner.

A leasehold property, on the other hand, will only confer rights to the owner leaseholder for the term of the lease, and the leaseholder will have to comply with the terms of that lease which may include paying a substantial rent. As is detailed in Chapter 9, the actual terms of the lease will have an impact on the price that should be paid for such a unit.

In simple terms, there are four different categories of hotel purchases:

- Vacant possession.
- Owned but operated by a tenant.
- Owned but operated by a manager.
- Owned but operated under a franchise agreement.

Vacant possession

Most lifestyle hotels are purchased as a vacant possession. Vacant possession, in the context of hotels, is different from most other types of property, for example houses and offices. If a house is sold with vacant possession it means that the property is empty, without a tenant or occupier of any type. However hotels are sold as an ongoing business, and so to sell it empty would adversely affect the trading.

As such, vacant possession for hotels means without a tenant or manager in place, (ie, without any legal operational encumbrances) but it will still be operational at the time of purchase, and the purchaser will benefit from future bookings. This also means that the purchaser will be discussing the staffing liabilities of the property, and will take over the stock (although this will be an additional sum on top of the agreed purchase price, calculated on the day of the transfer at cost).

Owned but operated by a tenant

A growing proportion of hotel investments across Europe are subject to operational leases. The owner of the property will either accept a premium at the start of the lease to allow the tenant to run the hotel, collect rent based on the agreement struck between the parties (this will be effected by the terms of the lease — see Chapter 9), or any combination of premium and rent.

The actual structure of the deal will generally depend on the needs and intentions of the parties. The tenant will then be responsible for the operation of the property and will employ all the staff themselves, undertake whatever repairs are specified within the lease at their own expense, and will be responsible for all things that are specified within the lease as the tenant's responsibilities.

Owned but operated by a manager

An important section of the hotel investment market, especially when involving US-branded hotels, comprise hotel investments being run by a hotel operator on a management contract. In effect, the owner is paying a manager to run the hotel on their behalf; the staff, the repairs and the operational costs are all the responsibility of the owner, albeit that the manager looks after the day-to-day running of the hotel for them.

The manager will charge a fee for his services, usually a management fee (based on revenue) and an incentive fee (based on profitability).

Operated under a franchise agreement

A franchise agreement is a contract that allows a hotelier/hotel owner to use the name (and usually have access to the central reservations system) of a brand in conjunction with the hotel. Such hotels will usually have minimum standards that have to be complied with, for example rooms sizes, facilities, safety and services, and the hotel will normally pay a fee based on bookings.

In return, the hotel will be branded (hopefully with a brand that helps them to generate more revenue than they would otherwise generate), and they should benefit from central marketing and bookings.

A hotel can be owned by one party who employs a manager (paying a management fee) and is subject to a franchise agreement (also paying a franchise fee).

An independent operator will need to persuade the hotel company of their ability to comply with the requirements of the brand (as well as their general good character) before they will be able to negotiate terms for a franchise agreement.

The various types of hotels that can be purchased

In simple terms, a hotel is a property that offers letting accommodation to paying guests. The level of service, the facilities and the price will vary in each hotel, but the common theme is that a transaction occurs; the guest contracts to stay at the hotel at a certain price and contracts to receive certain things in return, whether just somewhere to lay their head or dinner, bed and breakfast.

The type of business that the buyer wants to run will, to a large extent, influence the type of property that they should look at, and different types of hotel are outlined below. This is not an exclusive list as different categories are being created almost every day, with each operator striving to provide his own specific service, and not wishing to be labelled. As such, these terms are merely to provide guidance when looking at different property types.

Five-star hotel

Country house hotel

Five-star hotels

Five-star hotels are normally located in major city centres and will usually be fitted out, furnished, maintained and provide services to the very highest standards. Some five-star hotels are also "Trophy hotels" for example, The Ritz hotel in Paris, or the Shelbourne Hotel in Dublin, where any sale would most likely attract purchasers prepared to pay a premium to own such famous and distinguished properties.

The key to five-star hotels is in the superior service, in addition to the quality of furnishings and fittings, and traditionally the level of staffing at such a hotel will be higher than in other classes of hotel to ensure that the guest feels spoilt.

Country house hotels

Although difficult to define, country house hotels are usually attractive, older properties that have been fully modernised and maintained to a high standard, in keeping with their age. They may well have historical connections or be of historical importance.

Country house hotels will often be located in or close to places of historic importance or an area of outstanding beauty. They generally aim to provide a level of service very similar to that of a five-star hotel.

Boutique hotel

Mid-market hotel

Town houses/boutique hotels/bijou hotels

Usually located in major cities, boutique hotels will generally be relatively small, top-quality hotels aiming to give a personal and exclusive level of service. The hotel is likely to be full of individual character and is targeted at clientele that wish to stay in a highly individual property.

The "bijou" hotel has now established itself in the market. It provides high-quality accommodation with a personalised service, but with very limited additional service ie, virtually no food or beverage facilities.

Mid-market (three and four-star) hotels

These are generally purpose-built properties and are often located in or close to large towns and cities, or close to major transport termini or on motorways/major trunk roads.

Trade at such properties will be determined by demand generators in the hotel's specific location although it will usually be a mix of both corporate and leisure guests.

The quality of accommodation can vary substantially in this market segment, with the four-star hotels providing a generally higher level of service and facilities than the lower-rated hotels in this sector.

Spa hotel

Airport hotel

Spa hotels/destination spa

Since 2000, there has been rapid growth in the spa hotel market throughout Europe, although spa hotels have existed for over 100 years. These hotels are either located near natural spas or, more recently, the hotel offers purpose-built, modern spa and leisure facilities, and trades mainly from leisure guests.

The increase in disposable incomes in many countries and the demand for short breaks in Europe have led to the significant growth of this area.

Airport hotels

These are generally purpose-built hotels that are located close to major international airports. They tend to produce most of their bookings from people using the airport, whether people flying out early in the morning, landing late at night, or aircrew.

In actual fact, these are a sub-category of other hotels, and can also be defined as budget, three, four and five-star hotels, although the main defining factor in their categorisation is that the majority of their business is derived from their proximity to the airport.

Seaside hotel **Historic hotel**

Seaside hotels

This category covers a wide variety of different types and grades of hotel, from the large conference properties to the traditional small seaside boarding house.

With the exception of the larger hotels which aim mainly at the business and conference market, the majority of seaside hotels are smaller, independent or family-run properties relying on seasonal tourists and leisure trade.

Older and historic three-star hotels

This category embraces a variety of hotel types and locations, including hotels of character and former coaching inns. Many of these types of properties have more recent extensions forming additional bedroom accommodation at the rear.

Such hotels are often located in or near to historic towns, and generally have a mixed corporate and leisure trade. The number of bedrooms is usually less than 50 and often the food and beverage revenue may form a higher proportion of turnover than is the case for more modern commercial hotels.

Budget hotel

Independent hotel

Aparthotel

Budget hotels/lodges

These are a relatively new hotel concept, appearing in the UK in the mid-1980s and growing rapidly in the 1990s and 2000s. The first budget hotels/lodges to be widely developed were mostly located on motorways and other major roads.

Others have followed a similar approach whilst certain brands have developed hotels usually attached or adjacent to a public house as they do not provide food and beverage facilities within the hotel.

Budget hotels are one of the fastest growing segments in the UK, and already provide the highest proportion of hotel accommodation in France. They are now found in a wide variety of locations, including city centres, airports, roadside locations and out-of-town industrial and retail parks.

Apartotels

Since 1990, there has been enormous growth in the US of "apartotels" and since 2000 the concept has been steadily introduced into Europe.

An apartotel is a small flat or apartment that the guest will stay in, rather than a hotel room, and it will usually have a separate living area and kitchenette. The level of services provided will vary depending on the category of the apartotel, from full five-star service to more basic accommodation. They are generally located in principle commercial centres where guests have a longer stay requirement.

Small independently owned hotels, boarding houses, guest houses and bed and breakfast accommodation

These can be of almost any quality although they are generally small, one and two-star type properties. They are found throughout the country and do not form a homogeneous group, except that they are not easily included in earlier categories of hotels.

The main characteristic is the relatively low level of facilities and services provided, although meals are usually available.

Grading of hotels

Grading of hotels throughout Europe is notoriously inconsistent and a five-star property in some locations would be regarded as a lower four-star hotel in other countries. As such, several hotel companies also internally grade their hotels so that customers know what to expect from the property.

Branding can sometimes be useful as a guide of what the customer should expect; if the hotel company has rigid standards in terms of brand standards, then the customer should know exactly what facilities and level of comfort to expect. Unfortunately, this can fall down if standards are not sufficiently robust, at which point the branding can be almost worthless to the customer as a guide to the quality of the hotel.

Unlike most European countries and many other countries around the world, the UK does not have an official hotel grading standard. The most commonly known grading organisations are those of the AA (Automobile Association) and the RAC (Royal Automobile Club of Great Britain); the English, Welsh and Scottish Tourist Boards also have the crown system of grading.

Finding the right property

Finding the right property is the difficult part; almost every hotel will have significant differences and unless the potential purchaser knows exactly what they are after it will be difficult to find the right property.

As with estate agents, there are a number of specialist companies that sell hotels and if the buyer has contacted them they will pass on details of properties that meet the general requirements, albeit that the properties they show the buyer will also be passed onto other potential buyers who have the same requirements.

There are a number of reputable brokers that can be used to find the right sort of property to meet the purchaser's needs in "off-market" transactions, but even they will be unable to search for the type of property that the buyer is really after if they are uncertain of what it is that is required.

The buyer should consider questions such as:

- How big should the property be, with how many letting bedrooms?
- What type of hotel is required, with what type of guest profile?
- What standard of hotel is the buyer hoping to operate (whether RAC, AA or ETB rating, or more generic, like B&B or country house hotel)?
- Should the property have a bar and restaurant that can be run as a separate cost centre from the hotel, or should it merely be a adjunct to the letting accommodation?
- What living quarters will the buyer want/need?
- Does the property need to be in good condition, or is the buyer hoping to renovate and improve the trade, enabling them to put their own mark on the property?
- Does the buyer require space to expand?

There will be many other questions that will need to be considered when finding the right property, and it is important that the buyer considers all eventualities at an early stage to save wasting too much time and effort in searching and inspecting unsuitable properties.

Affordability

It is generally considered to be important (by the purchaser themselves, their advisors and their funders, although less so by the vendor) that the purchaser does not pay too much money for their desired property. Any overpayment could well eat into capital that could otherwise be better utilised to improve (or maintain) the property, or it could mean that interest charges on the purchase loan eat into a higher proportion of the earnings from the property.

From that perspective it is vital to work to a sensible budget, and then to work out a business plan to calculate how the business will work. This should not just be an exercise on raising finance from a bank, although it will be difficult to borrow money on a hotel without showing a sensible business plan. The whole exercise of going through the writing of the business plan should enable the purchaser to go through the whole thought process of the transaction.

The completed business plan should detail why it is being brought, why the price is reasonable, where the current trade is coming from, where the trade will come from in the future, the proposed pricing structure, sales and marketing plans as well as outlining staffing and resources plans. It should highlight the strengths and weaknesses of the current property operation, as well as the strengths and weaknesses of the prospective purchaser as an operator, going into operational risks involved with the project. That way, it will be possible to show the bank they are not looking at the project through rose-tinted glasses.

The plan should go into some detail on the customer base, the competition, changes to the trading environment, as well as detailing proposed repairs and improvement programmes.

The bank will want to see a cash flow analysis for the property, which must detail the proposed expenditure (purchase price and transactions costs, followed by future capital expenditure for any improvements or refurbishments) and a projected profit and loss account for the property down to EBITDA levels (earnings before interest, taxation, depreciation and amortisation). The plan should take into account the cost of interest on the loan and repayments to demonstrate that the debt will be affordable.

It is also useful to undertake a sensitivity analysis to show the impact that any problems with trading could have on the loan, which will help the bank be confident that the business will be able to support the proposed level of borrowing.

All of these items should be looked into carefully to ensure that the price that is being paid is sensible, affordable, and will enable the buyer to achieve what it is that is wanted from the purchase.

Finance

Most hotels will be purchased with a combination of bank finance and owner's equity; how much equity is required as a proportion of the whole purchase price will depend upon the circumstance of the purchaser.

In general, the higher the equity that you are prepared to invest in the project yourself the more keen banks will be to lend you the rest, as they are less at risk. This will usually reflect itself in more favourable loan terms. Other considerations will be the quality and desirability of the hotel, the pricing of the hotel, as well as your experience and credit worthiness.

It is beyond the remit of this book to discuss financial lending terms but there are many excellent mortgage brokers specialising in hotel funding who will be able to search out the very best terms available for you in your specific circumstances.

Reviewing the existing business

In Chapter 2, we outline in more detail how to assess the trading potential of a business. When a prospective purchaser is looking to buy a property to trade independently, it can sometimes be useful to visit the property many times and at various different times of the day.

Stay over unannounced as a guest to see what type of service the customers are receiving. Stay for breakfast; come back for lunch and during the early and late evening to see for yourself the dynamics that drive that particular business. Visit during the week and then also during the weekend. The more times the property is seen and the service experienced, the more opportunity there will be to assess the current trading and look for untapped potential in the property.

Also, it is useful to stay at the competition to see what they do well and badly, to review their cost structures and to see what sort of business they attract, in short to identify things that should ideally be done, or not done, when the property has been purchased.

A key question that needs to be assessed is how effective is the current manager — how much of the trade is generated by their ability and that could be lost if you do not have the same abilities? This is effectively the "personal goodwill" that is expected to leave the property and move with the present owner. This element of trading must be ignored when making assessments in your business plan, although it may be able to replace this business with improved trading that the purchaser's special abilities can bring to the business.

The importance of due diligence

However many times you visit the property and review what it happening and what is proposed, it is still essential that the purchaser commissions expert due diligence advice.

This book is mainly about putting a price on a hotel and valuation due diligence is obviously very important, whether to raise finance for the purchase or to provide peace of mind that the price is reasonable. Such advice should always be provided by an expert in hotel valuations.

There are many other types of expert advice that the purchaser may need to take, including:

1   Structural surveys — it may be that the property is old, in a state of potential disrepair, or in need of modernisation or refreshment. Sometimes it will be useful to have a full structural survey done, so that future plans for the property can be mapped out properly, with an accurate estimate of what costs might be incurred.

2   Financial — it is useful to have financial due diligence carried out on the property. The valuer will look through the accounts to assess the value of the property, but this will not be detailed analysis and cross-checking. Financial due diligence will provide some comfort that the stated accounts are accurate, and that forward booking are accurate, deposits are still held in place, that no tax liabilities exist (assuming a company purchase is being considered rather than a straight forward property transaction), and that everything is as the purchaser has been led to believe by the vendors. The financial advisor should also be in a position to start your tax planning straight away to ensure the correct structures are put in place to suit the purchaser's circumstances.

3   Legal — as with any property, it is essential to carry out legal due diligence. This can be confined to straightforward conveyance advice (it is freehold with no rights of way, good marketable title etc), or it can extend further into all the contracts involved with the purchase (staff employment contracts, future booking contracts etc).

The level of your requirement will depend upon the type of the property that is being purchased.

Summary

The potential purchaser should look carefully at the motivation for the purchase to ensure the most suitable property is purchased to meet your requirements. The type, location and ownership of the property should be suited to your specific requirements as much as possible to ensure that you have the best possible chance for success in operating the hotel.

The valuer will need to understand the dynamics behind potential purchasers, the demand and the resulting competition for such properties to accurately assess the property's value.

Understanding the Trading Potential of the Hotel

Introduction

This is probably the most important section of the book, as it is the key to getting the value correct and paying the right price for a hotel.

The value of a hotel (whether annual or capital) is based upon its potential trading, and nine times out of ten a purchaser/operator will buy (or rent) the property based upon the anticipated profit level it can sustain. For both the potential purchaser and the purchaser/valuer it is essential to know exactly how the property has been trading in the past to enable them to accurately assess how it is likely to trade in the future.

It is important to remember that the hotel business can be highly cyclical, with trading in many categories of the hotel highly dependant upon a number of outside events, including currency fluctuations, wars, interest rates, tourism cycles, terrorist acts as well as more general economic cycles.

When carrying out the hotel valuation, getting to the sustainable EBITDA (or divisible balance in the case of annual valuations) is usually carried out through five quite separate exercises:

1. The inspection.
2. The review of the competition.
3. The analysis of the historic trading accounts.
4. The interview with the general manager/financial manager.
5. Independent analysis of everything that has been seen and said throughout the process to determine how the hotel is likely to trade in the hands of the reasonably efficient operator (REO).

In assessing the purchase price that can be afforded (or carrying out a market valuation) it is important that any "personal goodwill" (extra trade that is generated by the owner) associated with the property is ignored while any "property goodwill" (goodwill that stays with the property rather than the owner) associated with the property is included. In assessing the potential performance of the REO the valuer is removing any element of "personal goodwill" from the equation.

The Valuation Information Paper number 6 called *The Capital and Rental Valuation of Hotels in the UK*, prepared by the Trading Related Valuation Group and published in March 2004 by the RICS stated that:

Detailed management accounts for an individual hotel, showing all items of income and expenditure, provide the base information for a Valuer. Particular matters to be considered include the following:

- whether the stated revenues and expenditures are in line with those considered achievable by a likely purchaser, or whether they are considered specific to the actual operator;

- companies with a number of outlets may not show outgoings such as marketing, training, accountancy, depreciation, cyclical repairs or head office management expenses in their individual unit management accounts;

- the accounts of an owner-occupier business may not show management salaries or directors' remuneration;

- depreciation policies vary, and it is important that the Valuer 'adds back' depreciation when assessing net maintainable profit;

- adjustments will need to be made to reflect the annual cost of repairing, maintaining and renewing the property and its fixtures, fittings, furniture, furnishings and equipment to an appropriate standard;

- hotels require refitting and re-equipping throughout their lifetime. It is important for the Valuer to take into consideration the time, income, cost and impact upon profits of such events, especially if major capital investment is likely to be required in the near future. In this case, an amount may be deducted from the valuation to reflect this.

The valuation paper then clarifies that:

The assessment of the trading potential of the hotel relies upon the Valuer's expertise and judgment.

The inspection

The inspection provides one of the best opportunities for the prospective purchaser (and valuer) to assess the value of the hotel.

There is much guidance available for purchaser/valuers on completing an inspection and it is not our intention here to talk about the duty of care offered to the client or the level of professionalism required to undertake a competent inspection as all of these requirements are automatically expected from a chartered surveyor.

Practice Statement 5.1 of the *Valuation & Appraisal Manual (Red Book)* summarises these requirements as follows:

Inspections and investigations must always be carried out to the extent necessary to produce a valuation which is professionally adequate for its purpose.

The *Red Book* also provides the following additional commentary:

Where a *property* is inspected the degree of on-site investigation that is appropriate will vary, depending upon the nature of the *property*, the purpose of the *valuation* and the *terms of engagement* agreed with the client.

2. A valuer meeting the criteria in PS 1.5 will be familiar with, if not expert on, many of the matters affecting either the type of *property* or the locality. Where a problem, or potential problem, that could impact on value is evident, from an *inspection* of the *property*, the immediate locality or from routine enquiries, an unconsidered *assumption* by the valuer that no such problem existed could be grossly misleading.

Understanding the Trading Potential of the Hotel

3. A client may request, or consent to, an *assumption* that no problems exist. If, following an inspection, the valuer considers that this is an *assumption* which would not be made by a prospective purchaser it becomes a *special assumption* and should be treated as such (see PS 2.2). However, these matters can rarely be disregarded completely and the discovery of adverse on-site factors, which may affect the *valuation*, should be drawn to the attention of the client before the *report* is issued.

4. Where it is agreed that inspections and investigations may be limited it is likely that the *valuation* will be on the basis of restricted information and PS 2.4 will apply.

5. Many matters which become apparent during the *inspection* may have an impact on the market's perception of the value of the *property*. These can include:

   (a) the characteristics of the surrounding area, and the availability of communications and facilities which affect value;

   (b) the characteristics of the *property*;

   (c) the dimensions, and areas, of the land and buildings;

   (d) the construction of any buildings and their approximate age;

   (e) the uses of the land and buildings;

   (f) the description of the accommodation;

   (g) the description of installations, amenities and services;

   (h) the fixtures, fittings and improvements;

   (i) any *plant & equipment* which would normally form an integral part of the building;

   (j) the apparent state of repair and condition;

   (k) environmental factors;

   (l) abnormal ground conditions, historic mining or quarrying, coastal erosion, flood risks, proximity of high-voltage electrical equipment;

   (m) contamination;

   (n) potentially hazardous or harmful substances in the ground or structures on it, for example, heavy metals, oils, solvents, poisons or pollutants that have been absorbed or integrated into the *property* and cannot be readily removed without invasive or specialist treatment, such as excavation to remove subsoil contaminated by a leaking underground tank or the presence of radon gas;

   (o) hazardous materials;

   (p) potentially harmful material present in a building or on land but which has not contaminated either. Such hazardous materials can be readily removed if the appropriate precautions and regulations are observed, for example, the removal of fuel (gas) from an underground tank or the removal of asbestos;

(q) deleterious materials;

(r) building materials that degrade with age, causing structural problems, for example high alumina cement, calcium chloride or woodwool shuttering;

(s) any physical restrictions on further development, if appropriate.

When reviewing a hotel, the inspection has five main functions over and above checking that the property actually exists:

1. *Check existing facilities*
   To check the facilities contained within the property and to accurately audit what is and is not included within the valuation.

   This may seem obvious, and would be undertaken by all potential purchasers and valuers in an inspection of any type of property. We have highlighted it here as it is not always simple when dealing with hotels.

   It could be that the property includes an annex with staff accommodation, that a conference building is held on a short lease (or different title than the main property), or that some of the plant and machinery (for example, telephone switchboxes and boilers) could be leased rather than owned and this may only become evident during the inspection.

   Complications can even arise as to how many rooms the hotel has, with some sources reporting in "rooms" and others in "keys". Therefore, care must be taken as a suite that has two bedrooms may be included within the letting inventory as "two rooms", but could at the same time be referred to as "one key".

2. *Assess the condition*
   To assess the size and condition of the property to ensure it is in good condition, or to enable the purchaser/valuer to make any necessary allowances for future capital expenditure.

   It is important that the condition of the property (with or without capital expenditure) ties in with the forecasted projections.

   For example a hotel may be able to achieve a RevPAR (average room rate multiplied by the occupancy rate, see definitions) of €65.00 if it is in good condition, but if the property has slightly more basic rooms and facilities it may be able to achieve a RevPAR of only €55.00. Without a full inspection, it is unlikely that the purchaser/valuer will be able to ascertain the probable trading level.

   Alternatively, the hotel may be projecting an ADR of €100, but it may contain too many single rooms to achieve such rates. Then again, it might be showing minimal conferencing revenue (at a lower level than would have been anticipated considering the listed facilities) and the reason for a poor trading record could become evident during the inspection.

3. *Analyse financial implications of the layout*
   To analyse any revenue and cost implications from the layout of the hotel.

   The design of a property can affect the efficiency of its operation and it is easier for an experienced purchaser/valuer to analyse how the property works during an inspection, rather than just from floor plans.

   For example, budget hotels strive to provide value-for-money accommodation to a relatively price-sensitive client base, with customers generally keen to stay in the most competitively priced property of a certain standard. As such, it can be important to minimise unnecessary costs, including an effective use of staff. A number of budget hotels have therefore "designed in" operational savings and these will have an impact on the EBITDA potential of the property.

A specific example of these types of designs is contained with the typical Express by Holiday Inn floor plate. The reception area (which needs to be manned day and night) is adjacent to the bar area (which needs less manning during the day). The idea is that during the day (and during quiet evenings) the reception staff can also provide a bar service, enabling the hotel to operate this facility without having to man both areas independently and thereby increasing potential revenue generation without incurring additional staffing costs.

The design and layout of the hotel's facilities can affect its revenue operation potential: meeting rooms without natural daylight can be difficult to let; bars and restaurants with separate off-street access could attract significant non-residential custom; if the vast majority of bedrooms have excellent views, then they could command enhanced rates.

4. *Look for treasure-hunting opportunities*

It is considered a truism in any business that the longer you have been with a specific organisation the less aware you are of other ways of doing business or, in other words, the less able you are to see unexplored opportunities. It can become common for a hotel to concentrate on what it is doing and to improve only the aspects of trading that have been determined to be the priorities, sometimes to the detriment of other potential trading opportunities.

One extreme example of this is illustrated clearly in an example when one hotelier purchased a property based upon the current and projected trading as provided at the time of purchase. When going through the property after the transfer was complete a fully functioning ballroom was discovered, which had been blocked off three years earlier through an operational decision of the existing hotelier. The projected accounts did not take into account any potential revenue that could be generated from this area and, as such, the purchaser had effectively underpaid for the property.

Other items that can be potentially "undervalued" in hotel valuations include unused or spare land that does not generate any income for the hotel and could be sold off without affecting the operation of the hotel.

Staff accommodation that is owned but no longer needed for the operation of the hotel can sometimes add significantly to the overall value of the hotel if, for example, the accommodation can be sold off independently or converted into revenue-generating guest accommodation. However, it is important to ensure that the true cost implications are taken into account: staffing costs may be lower because of the provision of staff housing and, as such, an adjustment to the accounts would have to be made to reflect this.

It is not the case that all staff accommodation is unimportant to the operation of a hotel and can therefore be sold off. If staff cannot be attracted to work at the hotel without the provision of staff accommodation, then it must be accounted for as part of the operation of the business rather than as a separate element of the hotel.

Other areas of potential treasure hunting include hotels that are not optimising the use of their facilities (such as, bedrooms could be converted into meeting rooms or free city-centre parking). All these need to be explored by the purchaser/valuer to assess what the REO could achieve.

It is part of the purchaser/valuer's duty to look for such treasure-hunting opportunities, and anything that is discovered, even if not something that is likely to be undertaken by an REO, should be mentioned in the valuation report.

5. *Review alternative uses*

The purchaser/valuer needs to assess the value of the hotel and review whether an alternative use would provide a higher value. This can be very important because a number of run-down hotels with poor trading records could have substantially enhanced values than those that a hotel operator would be prepared to pay for the property.

There are a number of common reasons why a hotel could have a higher alternative use value, including the following:

- It is poorly located for its use as a hotel (perhaps it has poorer access following development around the property).
- The property no longer meets the requirements of its marketplace (for example, non-ensuite bedrooms).
- The property has been left to deteriorate and it would cost too much to bring it back into a good enough condition to trade in its market.
- The property was poorly located when built and was only constructed as a planning condition of a larger mixed-use scheme.
- Other land uses in the area have increased at a faster rate than hotel land values.

When reviewing alternative use values the purchaser/valuer must be aware of the local planning regime and take into account the uncertainty of gaining planning permission for a new use, as well as the cost of demolition, conversion and rebuilding.

Limitations of the inspection

The problem with inspecting a hotel is that it is normally impossible to visit every bedroom during one inspection because a number of the bedrooms are likely to be occupied by guests at any one time. That said, it is not usually necessary to inspect every bedroom, although only the visiting purchaser/valuer will be able to make that determination. If the property is a 150-bedroom budget hotel and all of the rooms are of a similar size and condition, then less rooms will need to be seen than when it is a hotel with very few rooms of a similar style, size or quality, comparatively speaking.

It is important to bear in mind the objective of the inspection when determining how large a sample the purchaser/valuer needs to review. The purchaser/valuer needs to be able to review the historic trading and future projections accurately and to do this a reasonable cross-selection of bedroom accommodation will need to be viewed, including the best and worst rooms.

Sometimes it may not be in the operator/vendor's interest to show the least attractive or smallest rooms to the purchaser/valuer and they may try and hide certain areas; during the inspection it will normally become apparent to an experienced purchaser/valuer if this is happening and it is normal practice for the purchaser/valuer to ask to see specific rooms and require a much more comprehensive inspection.

Time of the inspection

The time that the inspection is carried out is likely to affect the number of bedrooms that a purchaser/valuer is able to see, with peak guest times (early morning and early evening) least desirable from the purchaser/valuer's perspective, as the least amount of rooms will be available for inspection.

However, if the majority of the revenue comes from the meeting rooms, then an early inspection time may be preferable so that this area of the hotel's accommodation can be fully inspected before it starts to be used. Alternatively, if most of the revenue comes from the bar or restaurant, the purchaser/valuer may wish to inspect the property while these areas are in operation, so that the trading pattern can be assessed.

Ideally, the purchaser/valuer will have the opportunity to stay overnight at the property, thereby experiencing the property as the guest would prior to formally inspecting the property and interviewing the general manager.

It is important that the inspection continues until the purchaser/valuer is happy that they have seen a good cross-section of the hotel's trading facilities and back-of-house areas (areas used by staff for the operation of the hotel, rather than being used by guests and directly generating revenue), as these will have an effect on operational costs and capital expenditure requirements.

It will be impossible to judge the correct EBITDA/divisible balance without having seen a satisfactory cross-section of the hotel and if the first inspection is not sufficient to provide an adequate sample, then a second inspection should be carried out.

Desktop valuations

This is one of the most contentious issues in hotel valuation — can you carry out an accurate hotel valuation on a "non-inspection" basis? There are various schools of thought on this but the issue must be settled by the ability of the purchaser/valuer to analyse the trading potential of the property and the requirements of the client.

The *Valuation & Appraisal Manual (Red Book)* provides general guidance on this point under Practice Statement 2.5:

> A revaluation without a re-inspection of property previously valued by the member, or firm, must not be undertaken unless the member is satisfied that there have been no material changes to the physical attributes of the property, or the nature of its location, since the last inspection.

The *Red Book* also provides the following commentary:

- It is recognised that clients may need the *valuations* of their *property* updated at regular intervals, and that reinspection on every occasion may be unnecessary. Provided that *members* have previously inspected the *property*, and the client has confirmed that no material changes to the physical attributes of the property and the area in which it is situated have occurred, a revaluation may be undertaken. The *terms of engagement* must state that this *assumption* will be made.

- The *valuer* must obtain from the client information of changes in rental income from investment *properties* and any other material changes to the non-physical attributes of each *property* such as lease terms, planning consents, statutory notices and so on.

- Where the client advises that there have been material changes, or if the *valuer* is otherwise aware that such changes may have taken place, the *member* must inspect the *property*. Irrespective of any changes to the *property* the interval between inspections is a matter for the professional judgement of the valuer who will, amongst other considerations, have regard to its type and location.

- The *valuer* may decide that it is inappropriate to undertake a *revaluation* without re-inspection because of material changes, the passage of time or other reasons. Even so the *valuer* may accept such an instruction providing the client confirms *in writing*, prior to the delivery of the *report*, that it is required solely for internal management purposes and no publication will be made to third parties. This position must be set out unequivocally in the *report* and state that the *report* must not be published.

It would be irresponsible of a purchaser/valuer to provide a desktop valuation if they were unable to be confident of the trading potential of the property. It is always possible to caveat a valuation to state that an inspection has not been carried out and that advice has been taken with regard to future capex requirements, but unless the purchaser/valuer can be confident enough of the actual asset, its facilities and its market position, then it is unlikely that an accurate valuation can to be carried out without visiting the property.

A number of the top hotel valuation practices appear to have a rough rule of thumb regarding such non-inspection valuations. Simply put, if the property has been fully inspected within the past three years, if no major capital expenditure has been undertaken, if the market has remained relatively stable over the period since the last inspection, and if all the information required is provided (including an interview with the general manager over the telephone), then a desktop valuation can be undertaken as it is possible to be relatively confident that an accurate assessment of potential trading can be made.

The local hotel market

It is vital for the purchaser/valuer to understand the specific parameters of the local hotel market if an accurate assessment of the sustainable EBITDA (or divisible balance) is to be possible. In that regard, the purchaser/valuer must find a way to assess the competitive supply of hotels in the area, the demand generators, and any proposed changes to the competitive supply.

The purchaser/valuer should also try and determine the usual performance in the area through historic benchmarking statistics as well as understanding the underlying nature of the hotel market.

The competitive supply of hotels

The first key question is "what constitutes competition for the subject hotel?", the answer to this will vary for every property. A hotel will provide competition for another hotel if they compete in some aspect, whether in food and beverage trade, leisure club guests, conference room demand or for rooms business.

That said, it is usual for a hotel's main competition to come from hotels of a similar quality in the local vicinity that compete for similar clientele. For example, in London the Four Seasons Canary Wharf is located less than one mile from the Ibis Docklands and they do not provide any competition for one another.

This example is perhaps slightly exaggerated because the difference in the quality of the products (the Four Seasons being a deluxe five-star property and the Ibis providing budget accommodation) and the differences in the markets in which they compete means that there is very little potential crossover of potential trade.

However, it is quite possible for the supply of budget hotels to have an adverse impact upon higher grade hotels and, in many areas, the mid-market hotels have seen their profitability decline after new openings of budget hotels in the same area.

Which hotels are deemed to be competition will normally be determined by the existing trading profile, although it is essential for potential purchasers/valuers to assess how the hotel could trade in

Understanding the Trading Potential of the Hotel

Four Seasons

Ibis Docklands

the future and review the competition accordingly. Competition will not just come from other hotels. If a high proportion of revenue comes from the restaurant and bar, then other restaurants and bars in the area need to be considered as direct competition. In addition, if the hotel has a fitness centre with an external membership, then other local health clubs will provide competition to the hotel.

Competition can also come from properties outside of the immediate area. For example, top-quality country house hotels that generate their own demand (ie, not location-based) are likely to compete with other such properties even when they are some distance from each other.

Another example is when a high proportion of revenue for the hotel comes from group tourism (coach tours for example), then other hotels that compete (or try to compete) in such markets should be reviewed to see whether the current revenue generated from this source is likely to continue unchanged.

It would be usual for a purchaser/valuer to inspect the most important competition, usually by requesting a guest show-round from the reception or meetings and events (M&E) staff. Hotels are regularly asked to see the quality of their bedrooms and meeting room facilities, and will normally try to accommodate such requests, although it must be borne in mind that it is usual to show only the most recently upgraded bedrooms and, as such, any rooms that are seen may not be truly typical of the hotel's bedroom supply.

It is also common for hotels to provide conference packs upon request that detail the meeting room facilities at the hotel, which can be useful for the purchaser/valuer.

A number of hotels also provide details of their rack rates (the published room tariff). It will be useful to note these, although the usefulness of these rates as an assessment as to what a guest is expected to pay will depend upon the pricing policy of the hotel. Some hotels will quote a rack rate and will expect to achieve that rate (this is more usual of the budget hotel groups who operate fixed pricing policies) but a large number of hotels publish one rack rate, but then offer substantially discounted rates that render the rack rate almost irrelevant.

Future competition

If the supply of hotels in a particular market segment changes, it is likely to influence the trading potential of the subject hotel.

If a competitive hotels closes down, this can lead to increased demand for bedrooms in other local hotels and so it is important to try and ascertain if any hotels are expected to shut down in the next few years. This can be quite difficult, especially if the competition is an unbranded, independent hotel. Some of the best ways to try and discover any potential closings is through speaking with the Local Authority, the general manager of the hotel that is being valued, and checking the local planning register to see if a change of use planning application has been submitted (or even granted) on a competitor hotel.

It is much easier for the purchaser/valuer to discover proposed new developments although determining which schemes are likely to come to fruition is more difficult, and will rely upon the experience of the purchaser/valuer and the quality of the information provided. Listed below are a number of methods to discover proposed developments in a local area that could prove useful:

- Review the Local Plan to discover the planning context for new hotels in the area.
- Review individual applications or consents for new hotels. This can sometimes be done online or through EGI's planning database, although care must be taken as to how up to date the information provided is.
- Discuss the local market with other property professionals.
- Discuss the local market with other hotel purchasers/valuers.
- Discuss the local market with the general manager.
- Review the local press for stories.
- Look at hotel company websites.
- Contact the local tourist board.
- Contact the local hotel association.

The demand profile and potential changes in demand

A hotel's performance will be influenced by the demand in an area and the nature of that demand. For example, hotels in Bristol in England will be influenced by a number of demand generators, including its position as a regional commercial centre for the South West of England, which will generate strong national, regional and local corporate demand for hotel bedrooms during the midweek period. It will also be strengthened through good tourist demand (from individual guests who pay higher-rate business and coach tours, which tend to pay slightly lower-rate business), which will strengthen some of the quieter corporate trading periods as well as generating bedroom demand over the weekends. However, other demand will also be experienced from people overnighting in the city on their way to other destinations, conferences organised by the city, events held in the city, overflow from events held at the Millennium Stadium in Cardiff and from the various sporting teams in the city.

To determine the demand generators in an area a purchaser/valuer will need to review both the business and the leisure profile of the location to determine where the trade is likely to come from. If the location is a relatively poor area with high unemployment this may lead you to believe that the corporate base for a hotel may be quite poor and so the emphasis behind the leisure trade will be more important.

Figure 2.1 RevPAR (£)

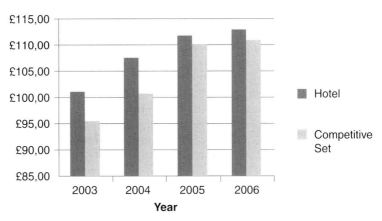

Source: Leisure Property Services

Unfortunately, it is not as easy as just counting the local businesses as certain industries generate more demand than others with, for example, banking generates more demand for hotel bedrooms (through training courses, relocations etc) than farming. It is also the case that different companies will require different hotel accommodation, so top accountancy firms may have more need for high-quality hotels (to accommodate their clients, partners etc) than a mechanic training school, where budget accommodation is more likely to be required.

Leisure tourism is no easier to determine accurately, as only certain countries and cities appear to accurately account for overnight visitors to an area. The popularity of a visitor attraction will not necessarily be converted into demand for hotels, as some attractions are not final destinations and do not therefore attract accommodation requirements as often as the visitor numbers would suggest. For example, Versailles Palace is one of the most popular tourist attractions in France but because it is so close to Paris, it is usually a day trip from the city or a stop-off on the way to another destination, rather than an end attraction.

Once the demand profile has been successfully identified, it will be important to look at the dynamics of the demand and review if this is likely to change. Changes can come from many areas, including an increase in new companies moving into the area, changes to the economic profile of the area, changes in transportation, inward investments etc.

Performance benchmarking

It should always be remembered that although a purchaser/valuer is looking to value the performance potential of a unique hotel, the requirement is to ascertain how the REO would be able to perform. If the hotel is underperforming or overperforming at a level that the hypothetical operator could not be expected to attain, then this must be ignored. It may be that the prospective purchaser would perform at an "atypical level". This should be mentioned in the valuation report, but will be ignored when the market value is assessed.

Valuation of Hotels for Investors

Table 2.1 Typical summary sheet

	2004		2005		2006	
No of Rooms	150		150		150	
Rooms Sold	37,482		37,914		38,270	
Rooms Available	54,750		54,750		54,750	
Occupancy (%)	68.5		69.3		69.9	
ADR	74.24		76.25		77.20	
RevPAR	50.82		52.80		53.96	
Growth (RevPAR) (%)			3.9		2.2	
Revenues (€000s)		(%)		(%)		(%)
Rooms	2,782.7	58.9	2,891.0	61.1	2,954.5	61.2
Food	*968.5*	*20.5*	*903.7*	*19.1*	*936.5*	*19.4*
Beverage	*581.1*	*12.3*	*529.9*	*11.2*	*555.2*	*11.5*
Total Food & Beverage	1,549.6	32.8	1,433.7	30.3	1,491.7	30.9
Room Hire	0.0	0.0	0.0	0.0	0.0	0.0
Leisure Club	292.9	6.2	298.1	6.3	289.7	6.0
Other Income	99.2	2.1	108.8	2.3	91.7	1.9
Total Revenue	**4,724.4**	**100.0**	**4,731.5**	**100.0**	**4,827.6**	**100.0**
Departmental Costs						
Rooms	768.1	16.2	795.0	16.8	824.3	17.1
Total Food & Beverage	942.2	20.0	860.2	18.2	893.5	18.5
Room Hire	0.0	0.0	0.0	0.0	0.0	0.0
Leisure Club	180.0	3.8	189.3	4.0	184.3	3.8
Other Income	34.2	0.7	37.0	0.8	34.8	0.7
Departmental Costs	1,924.5	40.7	1,881.5	39.8%	1,936.9	40.1
Departmental Profit						
Rooms	2,014.6	72.4	2,096.0	72.5	2,130.2	72.1
Total Food & Beverage	607.4	39.2	573.5	40.0	598.2	40.1
Room Hire	0.0	100.0	0.0	100.0	0.0	100.0
Leisure Club	112.8	38.5	108.8	36.5	105.4	36.4
Other Income	65.0	65.5	71.8	66.0	56.9	62.0
Total Departmental Profit	**2,799.8**	**59.3**	**2,850.0**	**60.2**	**2,890.6**	**59.9**
Undistributed Operating Expenses						
Administrative & General	387.4	8.2	397.4	8.4	410.3	8.5
Sales & Marketing	193.7	4.1	146.7	3.1	193.1	4.0
Property Operations & Maintenance	151.2	3.2	118.3	2.5	144.8	3.0
Utility Costs	153.5	3.3	153.8	3.3	156.9	3.3
Total Undistributed Expenses	**885.8**	**18.8**	**816.2**	**17.3**	**905.2**	**18.8**
Income Before Fixed Costs	**1,914.0**	**40.5**	**2,033.9**	**43.0**	**1,985.5**	**41.1**
Fixed Costs						
Reserve for Renewals	0.0	0.0	0.0	0.0	0.0	0.0
Property Taxes	198.4	4.2	203.5	4.3	207.6	4.3
Insurance	40.2	0.9	47.3	1.0	45.9	1.0
Management Fees	0.0	0.0	0.0	0.0	0.0	0.0
Head Office Costs	0.0	0.0	0.0	0.0	0.0	0.0
Rent	0.0	0.0	0.0	0.0	0.0	0.0
Other Costs	0.0	0.0	0.0	0.0	0.0	0.0
Total Fixed Costs	238.6	5.1	250.8	5.3	253.4	5.3
EBITDA	1,675.4	35.5	1,783.1	37.7	1,732.0	35.9

Source: Leisure Property Services

This does not mean, however, that the performance should be the average performance of all the potential operators for the unit. Some hotels will attract either particularly high-performing REOs, while others will attract those that do not perform at such high levels.

To be able to accurately ascertain how well (or poorly) the hotel is performing, it is useful to benchmark the hotel's performance against its competitive set. Recent legislation has made it illegal for hotels to discuss individual trading levels with one another on a formal basis, and so it is no longer possible to receive a list of Average Daily Rate (ADR) and occupancy information on a hotel-by-hotel basis for the competitive set. The valuer, however, should have enough immediate experience in the particular location and market segment (and indeed contacts) to know how that particular market set is performing.

More general information on performance levels is available from companies such as TRi, Deloittes and the Bench. They can provide generic location information (for example, all chain hotels in Moscow ran at an occupancy level of 69.2% at an ADR of €188.95 in 2006), or more detailed performance information (say, the average occupancy level of five specified mid-market hotels in Milan that can be used as a competitive set benchmark).

Although such benchmarking will not determine the occupancy or ADR level that the purchaser/valuer should attribute to the subject property, it is exceedingly useful in assessing the various underlying trends in that local market.

It is also beneficial for the client (whether they are a bank looking to lend on a property, a potential purchaser looking to buy a property, the board of directors in an annual review or an operator looking to take out a lease on the property) to see how the property has been comparing year-on-year with the local market.

Analysis of the accounts

The analysis of the historic accounts is an exceptionally good way to start to understand the property and the peculiarities of how it has historically traded. However, the purchaser/valuer must bear in mind that the historic accounts only show how that particular operator has been running the hotel, and that may not reflect the position of the REO.

A basic understanding of company accounts is required and it is not the intention of this book to go into the specifics of bookkeeping. However, it is important to draw your attention to the Uniform System of Accounts, which has been used for hotel accounting since 1926 and the 10th version was published in August 2006.

In a nutshell, the Uniform System of Accounts determines where specific revenue and costs should be accounted for in the profit and loss accounts, so that all hotels have a relatively comparable set of accounts for benchmarking.

For example, if one hotel accounts for its room profits after having deducted sales costs (which are usually attributed to Undistributed Operating Expenses under Sales & Marketing), then its profit margins will be lower than would be anticipated when looking at other similar hotels, which could confuse the purchaser/valuer when looking into the departmental profit margins.

Table 2.1 shows a typical summary sheet, which outlines the main headlines from the accounts. This summary would then usually be backed up by all the detailed accounts for the hotel.

The first part of the accounts details the statistical analysis of the performance of the hotel, working out the occupancy level, an Average Daily Rate (ADR) and the Revenue per Available Room (RevPAR) for the property. These statistics are very useful for benchmarking the hotel's performance against

Table 2.2 Profit and loss account for hypothetical hotel

		2006 (£)
Income		
	Accommodation	244,811
	Bar Drinks	1,289
	Breakfast	14,899
	Interest Receivable	68
	Telephone Parking	515
	Other	1,233
Total Revenue		**262,815**
Cost of Sales		(−£)
	Commissions Payable	10,865
	Direct NI	3,565
	Direct Wages	75,357
	Food & Provisions	5,882
	Laundry & Cleaning	10,311
	Purchases	491
Total Direct Costs		**106,471**
Expenses		(−£)
	Accountancy Fees	2,808
	Advertising & PR	889
	Bank Charges	399
	Credit Card Charges	1,965
	Depreciation	7,545
	Insurance	5,995
	Interest — Bank	7
	Light & Heat	6,525
	Car	8,250
	Director's Salary	52,000
	Rates	32,929
	Rent	50,000
	Repairs & Maintenance	17,889
	Reservations & Bookkeeping	2,150
	Service Charges	500
	Software	225
	Stationary & Printing	130
	Subscriptions	25
	Sundry	1,875
	Telephones & Fax	2,366
	Wages & Salaries	18,500
Total Expenses		**197,881**
Profit/Loss		**41,538**

Source: Leisure Property Services

other similar properties. However they are merely statistics generated from the detail behind Rooms revenue, which is detailed in the second section.

The second part of these summary accounts outlines the total revenue, breaking it down into the constituent departments that generated the revenue. In the detail behind the summary sheets the revenue stream will be further broken down, so for example the Food & Beverage revenue will usually be broken down into the various areas where it is generated. For example, the hotel may have earned 20% of its food revenue from the bar, 15% from room service, with the remaining 65% from the restaurant. However, this is likely to be further analysed so that the restaurant revenue can be seen to comprise 45% breakfast revenue, 40% evening revenue and 15% for lunchtime. The Rooms department may, for example, break down the revenue by industry segment, so that analysis can be made of where the most demand is being generated and what segments are the most lucrative in terms of ADR, or even by a geographic guest profile.

The third part of the accounts breaks down the departmental revenue into departmental profits, so, for example, the staffing costs in the restaurant along with the food costs will be deducted to show a departmental profit margin. Some hoteliers do not remove wages when calculating the departmental profits, which will lead to the profit margins looking much higher than would be expected; in this instance they are usually charged against undistributed costs as a Salaries & Wages cost. However, allocating the cost this way is not in strict accordance with the Uniform System of Accounts.

The fourth section is the Undistributed Operating Costs, which as the name suggests are costs that are accrued by the hotel that are not directly attributed to a hotel department. For example one of the sub-sections is Sales & Marketing — this is essential to the success of the hotel but cannot really be attributed to any one department, as good marketing may lead to a bedroom sale or a meeting room hire, but then the customer may also eat breakfast, go to the bar, spend some money in the shop, have a massage or use the telephone.

The last section is the Fixed Costs. These are costs that are born by the hotel whether or not it is operating efficiently, for example Insurance costs are unlikely to decrease just because the hotel is running at 40% occupancy. Other deductions include any outgoing rent, and head office costs or any other fixed costs that are not specifically accounted for elsewhere.

There is also usually an allowance for furniture, fittings & equipment (FF&E). The theory behind this is that to enable a hotel to maintain its current level of trading it needs to maintain the quality of the product and to do so it is usual practice to make an allowance for an FF&E reserve over and above the Repairs & Maintenance budget. That way the bedrooms can be systematically refurbished without needing to resort to capital expenditure. The amount that valuers will normally allow for this will be anywhere between 3% and 5% of turnover, depending on the quality of the property, its location and the amount the market normally adjusts for a property of this type to budget for the FF&E account.

The fixed costs also have one other cost centre — Management Costs. If the hotel has a management cost it will be deducted here, and indeed in the event of the property being subject to a management contract with a base management fee and an incentive fee then there will be two deductions.

However, a number of hotels are owner-operated and so do not deduct an allowance for the cost of management. In certain circumstances, it is appropriate for the purchaser/valuer to make a deduction to represent a notional management fee, for example if the property is run independently at the moment but would usually be run by a chain operator who would either charge a management cost, or re-charge some central head office costs (such as computer systems, human resources, accounting etc). In this case, it is essential that the purchaser/valuer looks at the accounts in detail to ensure that any costs that would not be accrued by a chain hotel are deducted so that the notional management fee is not "double-charging" the hotel.

Valuation of Hotels for Investors

Table 2.3 Amended Accounts to strip our "individual costs"

		2006 (£)
Income		
	Accommodation	244,811
	Bar Drinks	1,289
	Breakfast	14,899
	Interest Receivable	68
	Telephone Parking	515
	Other	1,233
Total Revenue		**262,815**
Cost of Sales		(–£)
	Commissions Payable	10,865
	Direct NI	3,565
	Direct Wages	**55,000**
	Food & Provisions	5,882
	Laundry & Cleaning	10,311
	Purchases	491
Total Direct Costs		**23,885**
Expenses		(–£)
	Accountancy Fees	2,808
	Advertising & PR	889
	Bank Charges	399
	Credit Card Charges	1,965
	Depreciation	–
	Insurance	5,995
	Interest — Bank	7
	Light & Heat	6,525
	Car	–
	Director's Salary	–
	Rates	32,929
	Rent	50,000
	Repairs & Maintenance	17,889
	Reservations & Bookkeeping	2,150
	Service Charges	500
	Software	225
	Stationary & Printing	130
	Subscriptions	25
	Sundry	1,875
	Telephones & Fax	2,366
	Wages & Salaries	18,500
Total Expenses		**145,176**
Profit/Loss		**141,524**

Source: Leisure Property Services

This will bring the hotel down to an EBITDA level, which is the standard unit that shows the level of profit that the hotel can make.

In many sets of accounts, depreciation and amortisation will be deducted leading to an EBIT level, but as each "hypothetical operator" will probably have different standards of depreciation and amortisation, they need to be added back in so that the profit margin is standard to all potential purchasers.

Other typical deductions that need adding back are interest costs, directors' fees, legal fees and other costs that are being deducted that result directly from the operation of the current owner exclusively and may be at different levels when the hotel is being operated by the REO.

Unfortunately, it is not always the case that the valuer will be presented with accounts that conform to the Uniform System of Accounts. It is considered good practice for the purchaser/valuer to reallocate the various costs into the appropriate parts of the Uniform System of Accounts layout, although it is not essential. What is essential is that the purchaser/valuer can understand what the accounts are showing, and analyse them enough to be able to ascertain the correct EBITDA levels.

A number of items that have been included as costs here to assess the tax position need to be added back to get to the sustainable EBITDA. According to the Uniform System of Accounts, the following costs would be struck out as shown in Table 2.3.

The *Valuation Information Paper* on the valuation of hotels in the UK states:

The purchaser/Valuer should be satisfied as to the accuracy and reliability of the trading information and/or projections supplied for the purpose of the valuation, and should critically examine the details against his or her knowledge of other hotels operating in similar trading circumstances.

If in doubt as to the accuracy of the underlying Assumptions supplied, verification should be recommended, with the valuation being referred back for review in the event of any material discrepancies arising from the verification process.

The accuracy and reliability of trading information is generally established by the auditing process. The Valuer will need to be aware of the relative reliability of audited and un-audited figures, and also consider whether or not the accounts are qualified or are a preliminary report.

This is an incredibly important point: the accuracy of the historic accounts is usually the starting off point for most prospective purchasers and valuers when assessing the future potential of the property. Any suspected inaccuracies or defects should be questioned, and indeed reported to the client to ensure everyone is aware of any lack of clarity that could impede the valuation process.

Analysis

The purchaser/valuer will find it useful to analyse the historic accounts that they have been provided with as it will be a starting point for discussions with the general manager and will enable the purchaser/valuer to understand the hotel operation. In the various sections below we provide an example of the sort of analysis a potential buyer and a valuer will undertake when assessing a property's trading potential.

Table 2.4 Part of a typical profit and loss account

	2004	2005	2006
No of Rooms	150	150	150
Rooms Sold	37,482	37,914	38,270
Rooms Available	54,900	54,750	54,750
Occupancy (%)	68.5	69.3	69.9
ADR	74.24	76.25	77.20
RevPAR	50.82	52.80	53.96
Growth (RevPAR) (%)	6.1	3.9	2.2

Source: Leisure Property Services

(i) Section 1, the operating statistics

As can be seen from Table 2.4, the bedroom numbers remained constant, which suggests that no extension works were carried out during the three years that are being analysed. The rooms available in 2004 were higher than the following years because it was a leap year.

The occupancy rate increased marginally in 2006 compared with 2005, but 2005 showed a marked improvement on 2004 (almost 2%). The purchaser/valuer would be asking themselves why this happened (was it an improvement in the general market, the result of improved demand at the hotel, a change of management, capex investment in the rooms, a change in the market mix etc, and more importantly would be trying to work out whether it is sustainable. Would the "hypothetical" operator be running at close to 70% occupancy or would 68.5% be more likely?

There was also some growth in the ADR experienced over the years, with a €2.01 increase in 2005 on the previous year (2.7%) and a €0.95 increase in 2006 compared with the previous year (1.2%). The purchaser/valuer will be considering whether the increases were purely down to inflation or market growth or whether these are due to the specifics of the actual property. The lower level of growth in 2006 compared with the previous year will need explaining — has the performance of the property peaked in the previous year and is now slowing down or was last year affected by something that will not recur in the following year?

Operators may decide to alter their business mix, either increasing occupancy levels at the expense of lower ADRs or by increasing the ADR and lowering the volume of business through the property. This is called "yielding" and it is the intention of every operator to try and optimise the rooms' yield that can be generated from the bedrooms by the best use of the mix of occupancy and ADR. There are arguments to say that an improved volume of business (at a lower rate) may lead to a bigger spin-off in secondary spending (with a higher volume of guests meaning a higher possibility of bar revenues increasing). There are also counter-arguments saying that an increased volume of business leads to higher servicing costs and reduces margins as well as increasing damage to the bedroom product through wear and tear.

It is the duty of the general manager to optimise EBITDA (or IBFC as many managers assume they can have limited impact on fixed costs) through efficient yielding. Equally, it is the duty of the purchaser/valuer to determine at what level the "REO" will operate at.

To check the yield generated by the rooms the hotel industry has come up with Revenue per Available Room (RevPAR), which is usually the occupancy level multiplied by the ADR. (In reality,

Figure 2.2 Actual RevPar, 2004–6

Figure 2.3 Growth in RevPAR, 2004–6

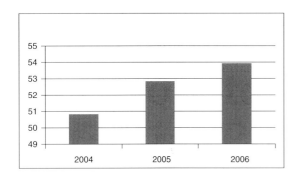

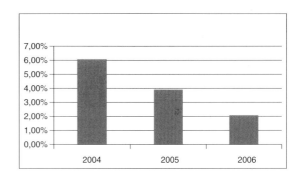

Source: Leisure Property Services

Source: Leisure Property Services

RevPAR should be the total revenue divided by the number of bedrooms, to truly reflect the revenue generated across all departments rather than just rooms but this is a standard measure which is a useful, if misnamed, tool).

The RevPAR tool only reflects half the story and the operator (and the purchaser/valuer) are also concerned with costs. So sometimes it can be useful to benchmark the gross operating profit per available room (GOPPAR) to determine the profitability of the hotel.

As can be seen from Figures 2.2 and 2.3, the RevPAR has been increasing each year, although the growth of rate has dropped each year. Once again, the purchaser/valuer will be looking at this to try and understand what has been happening and what is likely to happen in the future.

(ii) Revenues

The purchaser/valuer will look at the revenue mix of the property, once again trying to understand what has been happening historically to try and ascertain what is likely to happen in the future. Are there likely to be any key changes in the business mix, has a new facility opened, or has an old one closed, for example.

The detail behind the rooms' department accounts will probably analyse in greater detail where the demand has come from, with the bookings divided into market segments. The accounts will probably detail the total revenue generated, the occupancy and the ADR for each segment, as shown in Table 2.6.

Segmentation analysis will be extremely useful for the purchaser/valuer when they are looking at what is happening in the hotel, and careful use of segmentation can enable the hotel to improve its performance.

Such analysis is also useful as it can show if a hotel is too reliant upon any one segment. After the 9/11 terrorist attacks in New York, for example, a number of hotels across the world that had a high dependency on airline business found that they were either losing their business or that crew contracts were being renegotiated at lower rates, impacting quite substantially on their revenue generation. Any hotels that had a high proportion of such business found themselves badly affected, compared with other hotels that had a better mix of room demand generators.

Valuation of Hotels for Investors

Table 2.5 Part of a typical profit and loss account

	2004	(%)	2005	(%)	2006	(%)
Revenues ('000s)						
Rooms	2,782.7	58.9	2,891.0	61.1	2,954.5	61.2
Food	*413.0*	*20.5*	*400.3*	*19.1*	*413.3*	*19.4*
Beverage	*247.8*	*12.3*	*234.7*	*11.2*	*245.0*	*11.5*
Total Food & Beverage	660.8	32.8	635.1	30.3	658.2	30.9
Leisure Club	124.9	6.2	132.0	6.3	127.8	6.0
Other Income	42.3	2.1	48.2	2.3	40.5	1.9
Total Revenue	**4,724.4**	**100.0**	**4,731.5**	**100.0**	**4,827.6**	**100.0**

Source: Leisure Property Services

Table 2.6 Proportion of business (%)

	2006	2005
Rack	2.56	0.49
Corporate Agents	13.16	7.76
Corporate	34.67	39.98
Conference	6.20	4.90
Exhibition	2.55	4.08
Airline	15.87	15.63
Leisure Group Series	3.98	6.59
Frequent Independent Travellers	5.36	6.08
W/End Package (Full Rate)	3.33	4.07
Promotions	11.04	8.95
Other	1.27	1.46
Total Rooms	**100.00**	**100.00**

Source: Leisure Property Services

Figure 2.4 Rooms segmentation — proportion of business generated by each source (%)

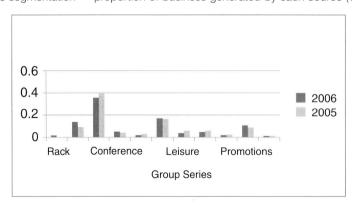

Source: Leisure Property Services

Other accounts may reveal a hotel's top clients, which will enable the purchaser/valuer to see how well spread the business is at the hotel. Once again, too much business coming from one source could be deemed "at risk" from a change in booking policy of that client in an economic downturn affecting that particular source of business and should be taken into account when undertaking the analysis of the accounts.

Looking at the accounts, the purchaser/valuer will see that although there was an improvement in rooms' revenue in 2005 on the previous year total revenue increased only slightly.

Food & Beverage (F&B) revenue was significantly down and the purchaser/valuer will want to understand why — was a new menu brought in?, were the meeting rooms taken off for refurbishment?, were the weddings down on the previous year?, was an outlet closed down?, was room service stopped? did a new restaurant open next door to the hotel?, has the allocation of breakfast revenue been altered? 2006 showed an improvement in the F&B revenue but it is still not as good as 2004 levels — why?

The purchaser/valuer will be wondering why the Leisure Club revenue dropped down in 2006 after having improved in 2005. This may simply be down to a drop in membership numbers, a decline in membership monthly payments, or possibly due to an inability to command the same joining fees. It could have come through the opening of new competition or because of an operational decision (perhaps the membership levels were too high and the use was adversely affecting the hotel guests' use of the facility).

The fluctuations of the "other revenue" department will also be analysed to understand what has changed year-on-year, so the purchaser/valuer can work out what is likely to happen in the future. It may be that a significant proportion of other revenue comes from leased out areas, and that these could be terminating in the near future or could be due to be reviewed, which could increase the rent received.

The purchaser/valuer will come up with questions for each revenue source which, hopefully, they will able to satisfy through the analysis of the accounts, the inspection and the interview with the general manager.

(iii) Departmental profits

As can be seen in Table 2.7, overall profits have remained relatively stable, with a slight improvement in 2005 followed by a slight downturn in 2006. The purchaser/valuer will look into why the Rooms' department is marginally down over the period, checking whether it is a straight forward increase in staff costs or whether there is something else behind the drop in profitability.

The improvement in Food & Beverage (F&B) margins will need to be analysed, but probably goes some way to explaining why the drop in F&B revenue has been accepted by the management. The purchaser/valuer will be trying to determine whether there is any possibility that the margin can be further enhanced, whether the revenue can be improved from this department without affecting the profitability, and whether the 40%+ profit margin is sustainable.

The purchaser/valuer will also be keen to understand what has caused the margins at the Leisure Club to deteriorate, especially since it does not seem to be related to the revenue stream (sometimes an improvement in revenue can lead to an improvement in margins as the departmental fixed costs become more efficiently used).

Table 2.7 Part of a typical profit and loss account

Departmental Profit	2004	(%)	2005	(%)	2006	(%)
Rooms	2,014.6	72.4	2,096.0	72.5	2,130.2	72.1
Total Food & Beverage	259.0	39.2	254.0	40.0	263.9	40.1
Leisure Club	48.1	38.5	48.2	36.5	46.5	36.4
Other Income	27.7	65.5	31.8	66.0	25.1	62.0
Total Departmental Profit	**2,349.5**	**49.7**	**2,430.0**	**51.4**	**2,465.7**	**51.1**

Source: Leisure Property Services

Table 2.8 Part of a typical profit and loss account

Undistributed Operating Expenses	2004	(%)	2005	(%)	2006	(%)
Administrative & General	165.2	8.2	176.1	8.4	181.1	8.5
Sales & Marketing	82.6	4.1	65.0	3.1	85.2	4.0
Property Operations & Maintenance	64.5	3.2	52.4	2.5	63.9	3.0
Utility Costs	65.5	3.3	68.1	3.3	69.2	3.3
Total Undistributed Expenses	**377.7**	**18.8**	**361.6**	**17.3**	**399.4**	**18.8**
Income Before Fixed Charges	**1,971.7**	**41.7**	**2,068.4**	**43.7**	**2,066.3**	**42.8**

Source: Leisure Property Services

(iv) Undistributed operating expenses

The fluctuations in the profitability of the "other revenue" will also have to be analysed to try and determine out what is the likely position for the future.

Once again, the purchaser/valuer will try and analyse why the various departmental sub sections have been altering over the past three years to try and work out where the costs are likely to sit in the future.

One question raised by these accounts are that 2005 undistributed costs were significantly lower than in either 2004 and 2006, and the purchaser/valuer will want to review these and see if they had an impact on revenue generation (specifically the drop in Sales & Marketing expenditure).

Another question would be the relatively static nature of the Utility Costs (energy costs) and how they have managed to remain a constant figure (in terms of proportion of total revenue) when the rest of the industry over this period is incurring generally higher costs.

Understanding the Trading Potential of the Hotel

Table 2.9 Part of a typical profit and loss account

	2004	(%)	2005	(%)	2006	(%)
Fixed Charges						
Reserve for Renewals	0.0	0.0	0.0	0.0	0.0	0.0
Property Taxes	84.6	4.2	90.1	4.3	91.6	4.3
Insurance	17.1	0.9	21.0	1.0	20.2	1.0
Management Fees	0.0	0.0	0.0	0.0	0.0	0.0
Head Office Costs	0.0	0.0	0.0	0.0	0.0	0.0
Rent	0.0	0.0	0.0	0.0	0.0	0.0
Other Costs	0.0	0.0	0.0	0.0	0.0	0.0
Total Fixed Charges	**101.7**	**5.1**	**111.1**	**5.3**	**111.8**	**5.3**
EBITDA	**1,870.0**	**39.6**	**1,957.4**	**41.4**	**1,954.5**	**40.5**

Source: Leisure Property Services

The purchaser/valuer will also need to check what is included under Property Operations & Maintenance to ensure that nothing has been deducted that should not be included under this section. Sometimes work that could be allocated as capital expenditure is deducted here, which would adversely affect the EBITDA of the property.

(v) Fixed costs

It is clear from Table 2.9 that there has not been a specific deduction for FF&E reserve, or indeed for a management fee or for head office costs. The purchaser/valuer will need to know if any head office recharges have been made elsewhere in the accounts, for example a central marketing cost under Sales & Marketing or a Human Resource recharge under Administrative & General.

It should be noted that the *Red Book* (PS 5.1) states that: "If detailed trading information is not made available, or if it is of insufficient quality, the purchaser/valuer must refer to this in the Report."

Once the purchaser/valuer has had a chance to review the accounts they should be in a position to have a meaningful discussion with the general manager or financial controller of the hotel.

The interview

The purpose of the interview with the General Manager (or the Financial Controller) of the hotel is to determine how the property has been trading historically, and where trading is likely to be in the future. Ideally, it should be a face-to-face interview (rather than over a telephone) and after the accounts have been received, as some obvious questions may arise through analysis of the accounts that may not otherwise be addressed automatically during the interview.

It is normal to send out a wishlist of information prior to inspecting the hotel that will hopefully be available for the purchaser/valuer prior to the inspection and these items are likely to provide a good basis for the interview questions.

The wishlist may include the following items:

1. Legal information
    - Title documentation or reports on title.
    - Details of ground leases or property leases.
    - Details of sub-leases.
    - Details of occupational licences or concessions.
    - Details of any problems with the title — re-privatisation problems etc.
    - Details of planning consents for each property.
    - Details of any unused planning consents.
    - Details of any unfulfilled planning conditions.
    - Copies of the operating licences — licence to sell alcohol, licence to operate as a hotel, opening certificate, fire certificate etc.

2. Property information
    - Size of the property.
    - Site plans.
    - Building plans — floor by floor.
    - Breakdown of bedrooms, Food & Beverage outlets etc.
    - Copies of any structural surveys.
    - Copies of any valuation reports.
    - Copies of any condition surveys, asbestos surveys etc.
    - Details of historic capital expenditure on the properties — (past three years).
    - Details of any proposed capital expenditure — this year and next year's budget.

3. Operational information
    - Full Profit & Loss accounts — three years minimum including occupancy levels and ADRs.
    - Full Profit & Loss budget for the current year and for the next three years (if possible).
    - Details of market segmentation and geographical analysis of the guest profile.
    - Details of contracts to supply bedrooms — historic and ongoing (corporate contracts/airline contracts etc).
    - Staff details — numbers of staff and contracts of employment, including pension details
    - Length of service and skills training (including languages) for each member of staff.
    - Sales & Marketing plans — this year and a budget for next year.
    - Details of Food & Beverage revenues — where it comes from, average cover spends, sleeper:diner ratios etc.
    - Details of the local markets — SWOT analysis of the local competitors.
    - Details of local trading — how does the property compete with its competitive set.

4. Other
    - Details of any outstanding complaints, legal claims etc.
    - Details of the tax positioning of the property (outstanding tax bills, capital allowances).

It would be impossible to compile a list of all the questions that the purchaser/valuer needs to ask during the interview as the answers that are received may lead to additional questions. Until a full understanding of the hotel's operation is gained, it is advisable to go through the interview in a logical way, perhaps following a similar approach each time to ensure that nothing is forgotten.

One such approach would be to review the trading in a similar approach to the layout of the accounts, ie, going through each departmental profit line first and then moving down to the EBITDA. For example, in a 100-bedroom conference hotel with leisure facilities the question structure might be similar to the following?

Bedrooms

Historic accounts — first the purchaser/valuer will try to understand the past performance of the hotel.

- What was the old occupancy level/ADR?
- Were the historic levels typical of the market?
- Was there anything particular that influenced the result (either positively or negatively)?
- What was the segmentation of the business like?
- Who were the top 10 clients for the hotel and what proportion of total revenue did they generate?
- Was there any specific geographic orientation to the clientele?
- Has any Capex been spent on the bedrooms recently, or do they need Capex?
- What profit margin is the Rooms department running at?
- Is housekeeping outsourced? Is the head housekeeper experienced and a long-term hotel employee?

Then the purchaser/valuer will try to understand the potential future trading profile for the property.

- What are the budgeted occupancy rates and ADR?
- Are they achievable? (Sometimes the budget is fixed by others, or is set at a very challenging rather than realistic level.)
- Are they reliant upon proposed Capex?
- How will they be achieved?
- How do they compare with the competition?
- Are they predicting any change in segmentation?
- Is there new supply coming online (or hotels closing down) that will effect future performance?
- Are there any major changes to the demand for such hotels in the area?
- What proportion of the budgeted trade is reliant upon one area/company/demand generator that could be at risk?
- Are there any changes that can be made to the cost structure of the department that will not adversely affect the quality of the operation?

The idea of these questions is to determine how the hotel Rooms department functioned, to gain an understanding of the demands and opportunities that exist for that segment of the hotel. It may be that each question leads to many more on a similar vein until the purchaser/valuer understands how that department works.

Then the purchaser/valuer would move on to Food & Beverage revenue.

- Where did the revenue come from (in terms of bar restaurant, room service, banqueting etc)?
- Is there any opportunity to change the revenue mix between the departments to improve both revenue generation and profitability?
- What was the breakdown in revenue between breakfast, lunch, afternoon tea, dinner etc?

- What were the guest conversion rates like (the proportion of guests that stay in the hotel to have dinner)?
- Who are the local competitors for evening meals and how do they compare with the hotel's offering?
- Is there much demand from non-resident guests for the bar or restaurant?
- What opportunities exist to expand the F&B revenue?
- How many weddings does the property cater for a year and can this number be improved?
- What is the volume of business like in the meeting rooms? Are day lets prevalent or is it easy to sell day delegate rates and residential packages to the guests?
- Are the rooms the right size for the market and do they provide the relevant level of flexibility?
- Are there any areas where the profit margins for the Food & Beverage are particularly low? Can these be improved?
- What level of service do the guests demand?

The purchaser/valuer should be aware of the different stresses that come from running the Food & Beverage operations in a hotel, and understanding the peculiarities of the cost control in this department will be essential, as almost every sub-department (whether the bar, the restaurant, the room service, the Meeting & Events) will have individual idiosyncrasies that will effect their potential profitability.

It may be that the REO would consider it sensible (and more profitable) to lease out the restaurant to a third party as a higher level of rent would be received that the hotel generates in way of departmental profit.

Then the purchaser/valuer would start to ask similar questions on the other revenue lines (including the Leisure Club revenue, Car Parking revenue, Telephone revenue and Minor Operating Department's revenue and working through their historic trading and future potential, so at the end of the conversation the purchaser/valuer should be confident of knowing precisely where they feel the average operator could expect to trade in terms of total revenue and departmental profits.

This questioning will then discuss the Undistributed Operating costs and Fixed Costs (although not all General Managers (GMs) seem interested in the Fixed Costs as, by their very nature, they cannot have much impact on these) until the historic details can be understood and the future predicted with some accuracy.

However sometimes it is not possible to interview the GM or Financial Controller through absence or the length of time for the interview is too short to get a feel for the operation of the hotel. Other times, the GM may be inexperienced or the purchaser/valuer may be expressly forbidden from carrying out an interview. In these instances, the purchaser/valuer will only be able to get a partial feel for the operation of the hotel, in which case a more detailed inspection and further local enquiries may be required to enable them to feel confident enough to produce a shadow profit and loss account for the hotel.

The *Valuation & Appraisal Manual* (*Red Book*) in PS 4.2 states that the valuer:

> must take reasonable steps to verify the information relied upon in the preparation of the valuation and, if not already agreed, clarify with the client any necessary Assumptions that will be relied upon.

At the end of the inspection, having gone through the accounts and undertaken the interview with the GM the purchaser/valuer should feel relatively confident about completing the shadow profit and loss account for the hotel, at least up until the IBFC line. It may be that the projections cannot be completed to the EBITDA line until the other enquiries have been completed.

Brand names

Should the property valuation include a brand name (or the rights to a trading name) then the potential of that name should be included in the property valuation. This value will form part of the "property goodwill" rather than the "personal goodwill" attributed to the property. If in the purchaser/valuer's opinion a prospective purchaser bidding for the property in the market might consider that the hotel, if sold without the brand name, may trade at a different level, then the difference in value should be identified.

Likewise, where a hotel without an existing brand or trading name is likely to be purchased by an operator with the benefit of a brand, then the increase or change in maintainable earnings and resultant value will need to be reflected by the purchaser/valuer.

Independent analysis of the information

The final stage of the process is vitally important. Hopefully, the purchaser/valuer will now have a good understanding of the individual dynamics of the hotel and will know what facilities they have, what the local market is like, where the hotel has been (in terms of trading) and where it hopes to go, and how it hopes to get there.

The purchaser/valuer will put all of this together and draw upon their knowledge of other hotels in the local area and of a similar type in other comparable locations to assess what the market would pay for this specific hotel, comparing the potential performance of the subject property against the potential performance of hotels that have recently sold, so that an accurate projection of trading for the reasonably efficient operator can be determined.

Summary

The potential purchaser and the valuer will need to completely understand the trading history of the hotel before they can accurately assess where the future trading potential can be expected to stabilise. It may be that the plan for the future hotel is substantially different from the existing operation (perhaps after substantial capital investment, or because of a change in the market dynamics in the location), but it is still important to know how the property has been trading, as that, to some degree, can impact upon the market perception of the hotel in the short term.

As such, it is important to find out as much as possible about the hotel through the inspection, analysis of the local market, analysis of the trading accounts and through discussions with the management of the hotel. It is vital that all of this information is fully assimilated so that an independent review of the data can be carried out in order to provide an independent assessment of the stabilised trading potential of the property.

Other Enquiries Required to Assess the Value of the Hotel

Introduction

There are a number of other enquiries that the valuer will have to undertake in order to accurately assess the value of the hotel. These include a review of the legal title to the property and all statutory matters affecting the property.

It should be born in mind that the valuer is looking for anything that could impact on the profitability of the hotel or anything that would cause a change in the multiple used to calculate value.

The *Valuation & Appraisal Manual* (*Red Book*) details other information that should be researched as part of the valuation process, which will include the following:

- planning (zoning) controls. Planning control will vary between states and the extent of the enquiries that need to be made will be governed by the valuer's knowledge of the area. The valuer must consider the nature of the *property*, the purposes of the *valuation*, the extent of the *property* and the size of the undertaking, in determining the extent to which the regulatory measures which can, or might, affect it should be investigated;

- the incidence of local, or state, *property* taxes;

- information on any substantial outgoings and running costs and the level of recovery from the occupier;

- information relating to any quotas imposed or other trading restrictions that may be made by the state in which the *property* is located;

- information revealed during the normal legal enquiry processes before a sale takes place.

Title and tenure

The value of the hotel will be affected by its tenure as the right to hold a property forever (freehold title) is more valuable than only being able to hold the property for another 20 years before it reverts back to the landlord, with or without the statutory protection.

If the property is leasehold there could also be a number of significant clauses in the lease that could affect value and it will be essential to review either the lease or a report on title produced by a lawyer.

In simple terms, anything in the lease that will deter someone from wanting to own the property could lead to a lowering in the value of the property, as the competition for the property is likely to be more restrained.

Key terms within the lease will include the following:

- The length of the lease (and when it expires).
- The user clause (if there is one) or any prohibitions on the use of the property. It is common for leases to specify a use for a property and to forbid certain uses. For example, the sale of alcohol or the use of the property for holding auctions are commonly excluded, both of which could affect the value if they were areas of trade that the hotel wished to become involved in.
- The demise of the property (what is included in the lease in terms of land and buildings) will directly impact on the value of the property.
- The rent payable and any rent review provisions. Any rent that needs to be paid to the landlord will be deducted under Fixed Costs in the Profit & Loss and the EBITDA will be reduced by this amount. Also, any provision for upward reviews of the rent needs to be assessed when determining the value of the property.
- The repairing liability. Although it is essential for the hotel to undertake necessary repairs so that the hotel is able to attract customers (which is a higher level of care than for most other property classes, which only need to be in good condition when they are being let, because hotels typically need to attract new tenants every night), an overly onerous repairing clause will adversely impact on the value of a property. Some leases require a certain proportion of turnover to be spent each year on the maintenance of the property, and if this percentage is too high it will detract from the value of the property.
- The insuring liability. Any unduly onerous provision will adversely affect the value of the property.
- The provision for alterations. Where the landlord has a right to prohibit alterations, whether absolutely or subject to certain provisions, this can have an impact upon the value of a property. Hotels are constantly evolving and unless they can adapt to the requirements of their clientele their trading will suffer, which will affect the value of the property.
- The alienation provisions (whether the property can be transferred in the event of a sale). Needless to say, the more restrictive the alienation clause the less potential market there is for the property, and therefore this could result in a lower value.
- Forfeiture provisions. Where the forfeiture provisions impact on the ability of the leaseholder to secure funding for the property it will have an adverse impact on the number of potential buyers for the property and, therefore, on the value.

There will be other key considerations contained within the clauses of the lease depending on where the property is located and the legal system that regulates such leases in that locality. However, it is not simply the difference between leasehold and freehold, as the quality of the title will also be important in determining the value of the asset. In this respect, reports on title are invaluable, as they have been prepared by experts and can be relied upon by the valuer.

For example, it may be difficult for a valuer to determine whether the boundary of the land on the title documents extends all the way to the public highway, and should the boundary not abut the public highway it could be that access to the hotel could be subject to control by a third party that owns the missing strip of land (sometimes known as a ransom strip). The lawyer should be able to advise on whether the hotel property connects with the adopted highways and services. It is not the intention here to discuss the impact of access issues, but they can be substantial.

The valuer will be expected to consider the ability of a purchaser to rectify any title defects, to secure an alternative access (and the cost of doing so) or the cost of an insurance indemnity policy, when looking at the impact such a defect could have on the value of the property.

Another typically difficult default in the title of a property may be in the provision of a restrictive covenant. Restrictive covenants are negative covenants and in English law can be difficult to expunge.

Restrictive covenants that are quite common in hotel documentation include the prohibition of specific uses (including forbidding the sale of alcohol, residential accommodation, car parking, auctions, and even for the use as a hotel) all of which could have a detrimental effect on the trading potential of the hotel and therefore its value. It could be that the hotel has a more valuable alternative use and so any prohibition of the land or buildings use for "other uses" would also detract from its value in that instance.

Other areas of defective title include where the quality of the title is not good and insurance needs to be taken into account. For example, in large areas of Eastern Europe there are problems with the title of property as the past landowners were forcibly evicted from the land either during the Second World War or during the following communist occupation. Their claims for repatriation of these lands has led to uncertainty in the quality of the legal title available for prospective purchasers.

Alternatively, only part of the land may have good clean title, with part of the property uncertain of its title position. In that case, the valuer will need to "take a view" on the importance of that part of the site to the overall operation. If it forms a fundamental part of the hotel the valuer will be expected to review the possibility of rectifying the title defect, taking out indemnity insurance, or replacing the affected part of the property on another part of the site. If it is part of the grounds, the impact of the "poor" title may be less significant.

Statutory enquiries

Property taxes

Another item that will need research (to assess the level of fixed costs of the property) will be the property taxes (or rates in the UK) and this can usually be found out from the Local Authority or Local Government collecting the tax.

In the UK, the rating liability changes each year, with a revaluation of the property every fifth year and the rates payable as a proportion of this rateable value being assessed on an annual basis. However, in the Ukraine there are no local property taxes and in Russia taxes are based upon a notional ground rent charge. Most countries throughout Europe use different methods of assessing the level of taxation, all based on very specific principles. In Chapter 10, we provide an outline on how property tax for hotels is assessed in the UK.

Planning and highways

The planning process varies across Europe but most systems are based upon similar principles, where land is zoned into specific uses and certain buildings and areas are more heavily protected than others in an attempt to preserve buildings of historical importance or areas of character.

It is important that a valuer knows the zoning of the land in which the property is built (along with any density restrictions where these apply) and the degree to which the property is "protected" or "listed" as these factors could impact upon any future redevelopment/extension of the property, or indeed effect the required Repairs & Maintenance expenditure for the hotel.

A valuer will need to look into the planning history of the subject property to ensure that it has consent for its current use, that it has consent for the building structure and that no outstanding planning conditions remain. Although there are statutory time limits that apply for presumptive changes of use and for retrospective planning consent, it would be usual that a hotel that either was trading outside its planning use or did not have consent for the structure would be worth less as a result of these defects.

A review of the planning history of the property will enable the valuer to see whether there are any outstanding planning decisions pending, or indeed valid planning consents that have not been undertaken.

If the valuer is being asked to consider the development potential of the hotel, then it can also be helpful to review the relationship with the Local Planning Authority. For example, if the hotel has previously tried to extend the bedroom accommodation only to be refused on specific grounds and if these grounds still apply it would be important to bring these to your clients attention.

It is also important to review the planning position of any sensitive sites near the property. For example, if the vacant lot next door has just been granted consent to build an incineration unit, that could adversely impact on the future trading potential of the hotel, and therefore its value.

The valuer also needs to be aware of any major planning changes in the area. For example, the provision of a new conference centre (which could increase business) or the provision of a new bypass (which could reduce the visibility of the hotel and therefore its business volume) and take any changes into account when looking at the value of the property.

The Valuation & Appraisal Manual (*Red Book*) provides the following guidance:

> The valuer needs to establish whether the property has the necessary statutory consents for the current buildings and use, and whether there are any policies or proposals by statutory authorities that could impact positively or adversely on the value. This information will often be readily available, but on other occasions delays or expense may be incurred in obtaining definitive information. The valuer should state what investigations are proposed, or what assumptions will be made, where verification of the information is impractical within the context of the valuation.

Statutory enforcers — operating licences

Each country has different requirements in relation to which operating licences or certificates that are required if the property is to be operated as a hotel. These may include fire certification, opening certificates, licences to sell alcohol, licences to extend usual opening hours in the restaurant or conference rooms, licences to carry out weddings and civil ceremonies and restaurant licences. As part of the due diligence process, the legal team and the valuer will be expected to determine whether these are all in place.

It is sometimes possible to make an assumption that these are in place but this assumption must be specifically communicated to the client (and in the event that this is for secured lending purposes this may not be acceptable) as any defect in the certification of the property can, in extreme cases, lead to the hotel closing down.

It is not the intention in this book to outline all the various operating licences required throughout the world, although anyone buying or valuing a hotel in a specific location should be aware of all legal requirements and should ensure that everything is in place.

In the UK, the requirements have changed recently in a number of key areas, which we outline below.

Fire certification

Fire safety regulations relating to non-domestic premises have recently been comprehensively reformed, with new regulations coming into effect on 1 October 2006. The new regulations, contained in the Regulatory Reform (Fire Safety) Order 2005, replace previous fire safety regulations, which were essentially regulated by two main pieces of legislation, the Fire Precautions Act 1971 and the Fire Precautions (Workplace) Regulations 1997.

Under the new regulations, responsibility for fire safety transfers from the Fire Authority to the "responsible person", ie, the employer, where there is one, the person in control of the premises for the carrying out of an undertaking, the owner, or any other person who exercises control over the premises.

Rather than the Fire Authority confirming what is required for compliance, the responsible person is required to undertake a fire-risk assessment, identifying the risks and putting in place appropriate fire precautions. The enforcement responsibility does, however, remain with the Fire Authority, which can inspect and serve alteration, enforcement and prohibition notices. Under the new regulations, current Fire Certificates cease to have any legal status.

This means that the valuer can no longer rely upon the provision of the fire certificate to assume that a "hypothetical purchaser" would consider it complied with the relevant requirements. The valuer now needs to be sure that the prospective owner would consider it safe (ie, does it have suitable fire access, is the fire alarm system adequate, does it have self-closing doors, fire doors, sprinklers in the kitchens, etc. Any defect in the "life safety" of the hotel will need to be reflected by the valuer by the way of additional capital expenditure (and even temporary closure during the works).

Justice's licence

With effect from 24 November 2005, a new licensing regime came into effect under the terms of The Licensing Act 2003. As part of the new regime, the responsibility for drinks licensing passed from local magistrates to Local Authorities. The aim of the legislation was to provide greater freedom and flexibility for the hospitality industry and to replace the old systems for regulating licensed premises with a single integrated system.

The Act requires all premises (where licensable activities are carried out) to obtain a licence and will help the local police to enforce the "licensing objectives".

A premises licence is required for any person carrying on a business that involves licensable activities (selling alcohol, providing late night entertainment). Once granted, the licence will last for an unlimited duration, although it can be revoked on appeal and will lapse if the holder becomes insolvent. When the business sells alcohol, it is mandatory that the sale of alcohol is authorised by a designated premises supervisor (who must be licensed and hold a personal licence).

It is important to note that the value of the hotel may depend upon the continuing ability to carry out licensable activities. If the premises licence lapses in an insolvency, then a key asset of the business could be lost.

Disability Discrimination Act

It is an offence not to provide equal opportunities for guests who are disabled to use a business' facilities, and this includes hotels. As such, Disability Discrimination Act (DDA) works should be carried out in all new hotels (and those carrying out significant improvements), enabling disabled guests to access all areas of the hotel (with the provision of relevant lifts and ramps) and have a number of bedrooms specifically adapted for their needs. Such adaptation may include wider doors to accommodate wheelchairs, wider bathroom access and handbars on baths or seats in shower areas, flashing lights and vibrating pillows for the hard of hearing or visually impaired, and different level facilities (such as a lower hanging rail in the wardrobe or eye hole in the door for guests in wheelchairs). The valuer will need to ensure that the hotel complies with such requirements or make suitable allowances for the work by way of capital expenditure.

Valuers should be aware of the potential liability on building owners or occupiers for work to comply with the DDA, especially where the property is used for the provision of goods or services to the public (for example, shops, leisure property and some offices).

Under the Disability Discrimination Act 1995 (DDA 1995), disability is given a wide definition incorporating most long-term and substantial impairments. Its scope is broad ranging and includes such diverse ailments as asthma, dyslexia, visual impairment and problems with mobility. Employers have a duty to make reasonable changes to practices and procedures within the workplace to enable disabled people to do their jobs. This may extend to making physical alterations to the workplace. These provisions came into force in 1996.

Part III of the Act, which came into effect on 1 October 2004, covers the provision of goods, services and facilities directly to the public. Since this date, service providers have had to take reasonable steps to remove or alter any feature that makes it impossible or unreasonably difficult for a disabled person to make use of the services.

The DDA 2005 amended the DDA 1995 to insert the disability equality duty, which is aimed at tackling systematic discrimination and ensuring that public authorities build disability equality into everything that they do. The DDA 2005 introduced new duties on property owners and managers, making it unlawful for landlords and managers to discriminate against a present or potential disabled tenant by failing, without justification, to comply with a duty to provide a reasonable adjustment for the disabled person.

Reasonable steps must be taken to provide an auxiliary aid or service to allow a disabled tenant to have enjoyment of the premises. A landlord may be obliged, for instance, to allow a wheelchair user access through an accessible alternative entrance not used by other tenants. The DDA 2005 does not require the landlord to make any physical alterations to the premises, however, a landlord must not unreasonably withhold consent to carry out alterations needed by the disabled occupier.

Statutory enforcers — health and safety

It is usual for most geographic jurisdictions to have a health and safety audit of hotels, whether checking the quality and cleanliness of the kitchens or whether they are likely to expose their guests (or staff) to legionnaires' disease; ideally, the valuer should make contact with such organisations to see whether the hotel has complied with all of its requirements or is about to be closed down through non-compliance.

Summary

There are many other factors that can affect the value of a hotel and will need to be researched, including the tenure of the property and the quality of the transferable title. The statutory framework of the local jurisdiction also needs to be reviewed, including the tax position, the planning position and whether all the relevant operating certificates and licenses have been issued and complied with. A defect in any of these can deter purchasers or lead to a lower sales price being achievable, and it is essential that the valuer is able to reflect this when undertaking the valuation.

Fundamental Principles Behind the Hotel Investment Market

Introduction

The purpose of this chapter is to provide a broad overview of the hotel investment market to help put this segment of the "buying" population into context.

It is important to understand the driving principles that guide investment acquisition decisions, as demand generated from investors can lead to increased competition in the acquisition process and potentially higher purchase prices.

What is an investment?

In simple terms, an investment is somewhere that money is placed to either protect the capital invested or to earn a return, or hopefully both. Property is one class of investment that can provide both an income (rent received) as well as capital growth. Other alternative investments include government bonds, savings accounts, art, fine wines and stocks and shares.

An investor will choose where to invest their money based upon the specific criteria of each investment and how is suits their needs. In general, it is true to say that the less risky the investment, the lower the returns received by the investor.

Property has been a popular investment for a long time. This is because it is considered to be an area where, in the long-term, capital values should rise, due to the limitations on supply and perceived long-term increases in demand. In the UK, for example, residential properties have seen huge rises in capital values over the past 100 years or so, led by the increased demand for properties (partly down to population increase, partly down to the breakdown of traditional family units and partly down the increasing availability of borrowing facilities). However, there has been a lesser increase in the supply of such properties (due to a number of factors not least various planning restrictions and a lack of available land).

Pension funds and other investment institutions have invested heavily in property in order to diversify their investment portfolios with the overall percentage of money invested in property varying year on year based upon the current thinking on the requirements for such investment vehicles. Traditionally, such investment tended to be limited to retail, commercial and industrial property investments, as they were seen as safer than other asset classes and there was a strong and measurable market.

In recent years, however, there has been a strong movement for diversification within property holdings and income-generating properties such as hotels have become increasingly popular with investors.

The range of hotel operators who are prepared to offer leases has increased substantially over the past 10 years. This increasing demand has led to a wide variety of hotel investment structures being created, from EBITDA-based leases, turnover leases, fixed RPI investment leases and management contract investments.

Who are the buyers and why are they buying?

One of the key questions is: who is behind this influx of money into the hotel investment sector since the mid-1990s and why? Is it that the investment world has finally woken up to the beauty of the hotel market or have the requirements of the investors changed, making it a much more attractive option than it use to be?

It is easy to categorise the hotel investors by corporate type:

- Private companies.
- Public companies:
    - pension, investment, and insurance funds;
    - governments and local authorities;
    - financial institutions.
- Partnerships.
- Private individuals.

In fact, this is a simple list of most of the different types of entities and individuals that exist in any investment market, not just the hotel market.

The issue, however, is not necessarily who or what the buyer is, but why they are seeking hotel investments? Broadly, the reasons are the traditional ones, capital growth and income, and hotels are rare in their ability to offer both.

To these generic, although appropriate, motives we must add some specifics, some of which are unique to the industry. The most obvious are hotel companies themselves, which want control over one of their raw materials, the hotels themselves (the other primary raw material being people). Then we must consider the other end of the scale, the individual, whose motives can include ego (as one banker describes it "a variant of the boys desire to own a pub and buy your friends drinks, just for slightly wealthier boys"), together with the decision to escape the rat race lifestyle, although this is far more commonly associated with owning and operating rather than investment. However, most investors are either looking for income or growth, or an acceptable combination of both.

A variant, but an important one, given the rise in popularity of hotels in recent years, is the broad-based property investor. Since the mid-1990s, hotel investment has become more and more attractive to companies that traditionally focused on commercial industrial and residential property.

Again, the motive must be examined. In the early cases it was an entrepreneurial assessment of an undervalued asset class that merited exploitation. The likes of Land Securities, London and Regional,

Heron and Marcol all utilised their real estate knowledge to create capital growth from hotel real estate while benefiting from higher yields. Mostly, they understood the increased risk associated with these investment decisions and these formed part of a broad investment policy, with government-let offices at one end, and luxury hotels at the other.

However, as yields have sharpened (in this context a lower yield means a higher price or a longer period before the initial investment is paid back), and since the new millennium interest rates across Europe have risen, some of the new investors have simply arrived in the sector chasing yield.

The culture of leverage has played its part. Traditional bankers used to say that the right amount of leverage was around 60% by value, and two times interest cover from earnings. Low interest rates, private equity, opportunity funds, and ultimately the increasingly distant memory of the last recession, have all contributed to increasing leverage in all markets, in turn leading to and combining with the wall of capital that drove leverages in excess of 85% in late 2006 for prime hotel investments.

Falling or sharpening property yields have had the effect of making debt less serviceable, and yet the leverage issue has forced debt levels up, not down, compounding the potential problems for hotel lenders.

This has resulted in investors seeking greater yield than has been available in their core real estate sectors (typically commercial - offices and retail), which has meant accepting higher levels of risk.

This has manifested itself in three main ways: the compression of secondary yields (and at a faster rate than primary property), the move into riskier transaction structures, and movements into investment classes with greater yield.

Primarily these have been industrial and hotels. One result is a wave of new investors seeking hotels not for their underlying business, but their combinations of yield and income.

The risk, and that underlying business that makes them what they are is a by-product, but certainly one that some investors have found quite pleasing once they took a look around their new assets.

What types of hotel investments are available?

Ground leases

The simplest and longest running type of hotel investment is the ground lease. Typically the landlord will collect a rent from the operator who will have invested their money into building the property, and the lease will be for a long period (typically 99 years as a minimum, although sometimes extending to as long as 999 years).

The rental stream can vary, from a peppercorn (a notional rental amount that is not usually ever demanded or paid), to a fixed annual sum throughout the term, a rent that is reviewed periodically (or annually) based upon the rental value of the site, or to a proportion of turnover.

It has latterly become common for good sites to be leased to operators on a turnover basis. This ensures that the landlord benefits from the investment while, at the same time, not unduly penalising the operator when trade declines. This flexibility, which usually results over the period of the lease in the investor receiving a higher rent, provides the operational flexibility required by the tenant to ensure that they can ride out any difficult markets, and has therefore come into favour on all sides.

Typically, ground rents are very secure (although the income stream from a turnover lease is uncertain and therefore not viewed as quite so secure) and therefore the returns tend to be lower than for other investment types.

Fixed leases

Hotels have also been let as fully fitted hotels, as a shell hotel with no fitting out, or anywhere in between, to operators on traditional leases, with rent reviews at varying periods of time (commonly reviewed annually, 5 yearly, 7 yearly, 14 yearly or 21 yearly).

The variety of this type of lease is almost infinite. It is common to see fixed rents granted with a rent specified for each year throughout the whole term of the lease, fixed rents annually reviewed to RPI, fixed rents with reviews every five years to RPI, fixed rents linked to the market rent (of hotels and other classes of property), fixed rents linked to a proportion of market rent or even fixed rents with review to a proportion of the capital value of the hotel.

A lease with a known rental stream throughout the whole term is considered to be more secure by investors (assuming identical levels of rent, lease terms, quality of tenant and all the other factors discussed in detail in the chapter regarding the choice of investment yields) than one that is linked to market rent.

Annual reviews are also considered more desirable (and therefore produce prima fascie lower returns) because there is a lower level of perceived risk than those that are reviewed less frequently.

Performance-based leases

There are also varying categories of performance-based leases, from those that pay a proportion of the hotel's net profit (EBITDA) to the landlord, to those that have an element of guaranteed income and an additional element based upon the level at which an agreed percentage of turnover/EBITDA exceeds that guaranteed element of rent.

Once again, the more secure and the easier it is to collect the rent is, the lower the returns are likely to be. EBITDA-based leases where the landlord is reliant upon the successful trading of the hotel transfer a high proportion of the "operational risk" across to the investor and, as such, the investor is likely to demand a higher return to compensate for this additional risk.

Management contracts

A management contract is an unusual investment for many types of investors as it transfers most of the "operational risk" across to the investor at the same time as transferring the property risks as well. The investor behind a management contract will be employing the operator to manage the property on their behalf, meaning that the staff are all employees of the investor (along with all the relevant employment regulations and legislative issues this raises), the repairing liability for the property is squarely on the shoulders of the investor, and if the manager (or management company) is not very good, then the investor's returns will be adversely effected.

Valuation methodology

The simplistic valuation methodology for the investment differs slightly from a vacant possession valuation, insofar as the investor's exit route has to be factored in, as this is a very real consideration for most investors when they consider the overall cost of an asset. It can be defined as:

$$\left[\frac{\text{Rent}}{\text{income}} \times \text{multiple} \right] - \text{transaction costs} = \text{value}$$

The transaction costs reflect the cost to the investor of realising the proceeds of the transaction and will usually include legal fees and agents fees (plus any relevant taxes on these fees, for example VAT in certain European countries) and any property transfer taxes.

Summary

The variety of different investment structures available in the hotel market has enhanced the appeal of the sector to mainstream investors; the differing types of properties within the hotel industry with their different fundamental principles allow for a diversification of risk, with certain sectors considered recession proof and other sectors benefiting directly from any growth in the economic climate as a whole.

The innovative structures that have been created have allowed investors to benefit from good performance while not penalising operators when times are hard, enabling a true partnership to evolve, which has encouraged all parties to want to be involved in the sector.

The Purpose of the Valuation and Statutory Guidance

Introduction

There are various "professional guidelines" that valuers must adopt and these are outlined mainly in the *Appraisal & Valuation Manual* (known as the *Red Book*).

In *Zubaida* v *Hargreaves* [1995], the so-called "Bolam test" was applied to valuers: When assessing negligence the issue is now universally agreed to be whether the defendant has acted in accordance with practices that are regarded as acceptable by a respectable body of opinion in his profession. If the defendant can show that the valuation process is in accord with that of such a respectable body of opinion, he will, broadly speaking, not be found to have been negligent. By the same token, however, if a respectable body of such opinion considers his conduct to have been negligent, that places him under a burden to offset or counter that opinion.

The purpose of the valuation does, to some extent, determine the appropriate methodology that needs to be adopted when valuing a hotel. However, the purpose of the valuation should not lead to differing results if the requirement is to determine Market Value.

In *Singer & Friedlander Ltd* v *John D Wood & Co* [1977], Justice Watkins stated that:

> The way in which a Valuer should conduct himself so as to fulfil his duty to a merchant bank, or any other body or person, varies according to the complexity or otherwise of the task which confronts him. In some instances the necessary inquiries and other investigations preceding a valuation need only be on a modest scale. In others a study of the problem needs to be in greater depth, involving much detailed and painstaking inquiries at many sources of information. In every case the Valuer, having gathered all the vital information, is expected to be sufficiently skilful so as to enable himself to interpret the facts, to make indispensable assumptions and to employ a well-practised professional method of reaching a conclusion; and it may be to check that conclusion with another reached as the result of the use of a second well-practised method. In every case the Valuer must not only be versed in the value of land throughout the country in a general way, but he must inform himself adequately of the market trends and be very sensitive to them with particular regard for the locality in which the land he values lies. Whatever conclusion is reached, it must be without consideration for the purpose for which it is required. By this I mean that a valuation must reflect the honest opinion of the Valuer of the true market value of the land at the relevant time, no matter why or by whom it is required, be it by merchant bank, land developer or prospective builder. So the expression, for example, "for loan purposes" used in a letter

setting out a valuation should be descriptive only of the reason why the valuation is required and not as an indication that were the valuation required for some other purpose a different value would be provided by the Valuer to he who seeks the valuation. It might, however, be an indication that the Valuer, knowing the borrowing of money was behind the request for valuation, acted with even more care than usual to try to be as accurate as possible.

In *Banque Bruxelles Lambert SA* v *Eagle Star Insurance Co Ltd* [1995], it was stated that:

> A Valuer's duty ... is to take reasonable care to give a reliable and informed opinion on the open market value of the land in question at the date of valuation ... In the absence of special instructions, it is no part of a Valuer's duty to advise on future movements in property prices; his valuation is not sought to protect the lender against a future decline in property prices.
>
> In the absence of special conditions the Valuer's duty to the Lender is the same: to take reasonable care to give a reliable and informed opinion on the open market value of the land in question at the date of valuation. In the ordinary way the Valuer does not warrant that the land would fetch on the open market the value he puts on it, any more than a medical practitioner warrants that he will cure a patient of illness. In each case the duty is to exercise a reasonable standard of professional care in the circumstances, no more and no less.

Introduction to the *Red Book*

The Royal Institution of Chartered Surveyors (RICS) has published Appraisal and Valuation Standards since 1974 (known as the *Red Book*). At first, they only applied to valuations incorporated in published accounts, but since the mid-1990s have applied to most types of valuations. Since 1991, compliance with the Practice Statements within the *Red Book* has been mandatory for members of RICS. The sixth edition was published in November 2007.

The RICS states that:

> The purpose of these standards is to ensure that *valuations* produced by members achieve high standards of integrity, clarity and objectivity, and are reported in accordance with recognised bases that are appropriate for the purpose.

With regard to trading-related valuations (which includes hotels), the RICS has produced a guidance note as an appendix to the mandatory part of the book, providing guidance as to how to carry out such valuations.

It is useful to review this guidance note in full (GN1), as the valuer should always seek to follow this guidance unless there is an overwhelming reason not to (for example, it would conflict with market practice and would therefore provide an inaccurate assessment of the value of the property).

GN1 Specialized Trading Property valuations and goodwill

1.1 The commentary to PS 3.2 indicates that special consideration must be given to the application of *Market Value* to certain categories of *property* that are normally bought and sold on the basis of their trading potential. Examples of this type of *property* include hotels, bars, restaurants, theatres or cinemas, fuel stations, and care homes. The essential characteristics of *properties* that are normally sold on the basis of their trading potential is that they are designed, or adapted, for a specific use and that ownership of the *property* normally passes with the sale of the business as an operational entity.

1.2 This guidance note is restricted to *trade related property valuations*. It considers the additional criteria that need to be considered by the valuer in these cases but does not concern itself with methods of *valuation*, which will vary depending upon the *property* to be valued.

2. **Terms used in this guidance note**

2.1 It is essential that the valuer recognises, and understands, that the terms used in this guidance note may have different meanings when used by other professional advisers. In particular the term '*property*' in valuation terms is defined in the glossary whereas in other contexts it includes intangible assets as well as intellectual *property*. Unless stated otherwise the term '*property*' is used in this guidance note as defined in the glossary. In this guidance note the following terms will be used as defined in the following paragraphs.

2.2 **Trade related** *property*
Property with trading potential, such as hotels, fuel stations, restaurants, or the like, the Market Value of which may include assets other than land and buildings alone. These properties are commonly sold in the market as operating assets and with regard to their trading potential. Also called property with trading potential.

2.3 **Valuation of the operational entity**
The assessment of the value of the operational entity will usually include:

- the legal interest in the land and buildings;
- the *plant & equipment*, trade fixtures, fittings, furniture, furnishings and equipment;
- the market's perception of the trading potential, excluding *personal goodwill*, together with an assumed ability to obtain/renew existing licences, consents, certificates and permits; and
- the benefit of any transferable licences, consents, certificates and permits.

Consumables and stock in trade are normally excluded.

2.4 **Reasonably efficient operator**
A market-based concept whereby a potential purchaser, and thus the valuer, estimates the maintainable level of trade and future profitability that can be achieved by a competent operator of a business conducted on the premises, acting in an efficient manner. The concept involves the trading potential rather than the actual level of trade under the existing ownership so it excludes *personal goodwill*.

2.5 **Goodwill**
Transferable Goodwill. That intangible asset that arises as a result of property-specific name and reputation, customer patronage, location and similar factors, which generate economic benefits. It is inherent to the specialized trading property and will transfer to a new owner on sale.

2.5.1 In previous RICS guidance *transferable goodwill* has also been referred to as inherent goodwill. In the valuation context these terms are identical. Other professional advisers, may use other definitions of types of goodwill, such as 'free goodwill', which do not have any bearing on the *valuation of property*. In relation to a *trade related property valuation*, goodwill is either capable of transfer with the *property* interest or it is not. Personal goodwill. The value of profit generated over and above market expectations which would be extinguished upon sale of the specialized trading property, together with those financial factors related specifically to the current operator of the business, such as taxation, depreciation policy, borrowing costs and the capital invested in the business.

2.5.2 *Personal goodwill,* where it can be identified, remains with the operator and, in principle, may be transferred to another *property*. The comment that the *personal goodwill* would be 'extinguished' refers to that element of *personal goodwill* to be excluded from the consideration of the maintainable level of trade of the subject property and not the *personal goodwill* itself which is an intangible asset.

3. **Valuation assumptions**

3.1 *Trade related property* will usually be valued subject to specific *assumptions*.

3.2 Where the *property* is trading and the trade is expected to continue a typical *assumption* would be:
Market Value as a fully-equipped operational entity having regard to trading potential.

3.3 Where the *property* is empty either through cessation of trade or it is a new *property* with no existing trade to transfer different *assumptions* are made. For example, an empty *property* may even have been stripped of all or much of its trade equipment, (or a new *property* may not have the trade equipment installed) but it could still be valued having regard to its trading potential.

3.4 Also the closure of a business, and the removal of some, or all, of the trade equipment, may have a significant effect on the value of the *property*. It will, therefore, often be appropriate to express the value on the basis of one or more *special assumptions*, as well as on a basis reflecting the status quo. This is often a requirement when advising a lender as to the value of *trade related property* for loan security purposes. It does not follow that the difference between this *special assumption* and the value reflecting the status quo represents the value of *transferable goodwill*, and valuers should not indicate any such apportionment. For example, the differences could reflect the cost and time involved in removing the fixtures and purchasing new equipment. Other examples of *special assumptions* are given in appendix 2.3.

3.5 In these cases the typical *assumption* will be:
Market Value of the empty *property* having regard to trading potential [subject to the following special assumptions ...]

4. **The valuation approach**

4.1 The *valuation* of a *trade related property* necessarily assumes that the transaction will be of the *property* interest, together with all the equipment required to continue operating the business (if it is assumed to be 'fully-equipped'). In this context equipment includes *plant & equipment*, fixtures, fittings and furnishings. However, care must be taken because this *assumption* does not necessarily mean that all such equipment is to be included in the *valuation*. For example, in the case of a *valuation* in connection with a proposed transaction or for security purposes some equipment may be owned by third parties and would therefore not form part of the interest being valued. In order to avoid misunderstanding the valuer should always establish what is to be included in the *valuation* when settling the *terms of engagement*, and make this clear in the *report*.

4.2 Where *assets* that are essential to the running of the entity are either owned separately from the land and buildings, or are subject to separate finance leases or charges, an *assumption* may need to be made that the owners or beneficiaries of any charge would consent to the transfer of the *asset* as part of a sale of the *operational entity*. If it is not certain that such an *assumption* could be made, the valuer must consider carefully, and comment in the *report* on, the potential impact on the *valuation* that would be caused by the lack of availability of those *assets* to anyone purchasing the operation.

4.3 The valuer needs to be aware of the distinction between the *Market Value* of an operational entity, and the value to the particular operator (its *worth* to that operator). The operator will derive *worth* from the current and potential net profits from the business operating in the chosen format. While the present operator will be one potential bidder in the market, the valuer will need to understand the requirements and the achievable profits of all other potential bidders, and the dynamics of the open market to come to an opinion of value.

4.4 A *valuation* on the basis of *Market Value* should only reflect the *transferable goodwill* that relates to the trading potential of the *property*. The *valuation* should exclude any *personal goodwill* to the present owner or operator which would not be passed to a purchaser of the *property*. For example an operator may attract significant custom (*personal goodwill*) to a specific *property* as a result of their celebrity status, or a corporate operator may attract significant custom as a result of their brand identity or that of a brand under licence. Such additional custom would move with that person, or company to a new location managed by them but would not transfer on the sale of the *property* to any purchaser. It may even be terminated by a licensor.

Assessing the potential

4.5 *Trade related properties* are considered as individual trading concerns and typically are valued on the basis of their potential earnings before interest, taxes, depreciation and amortisation (EBITDA) on the *assumption* that there will be a continuation of trading. The EBITDA of the actual business will often be different from the valuer's EBITDA which is based on an assessment of the market's perception of the potential earnings.

4.6 The task of the valuer is to assess the fair maintainable level of trade and future profitability that could be achieved by a *reasonably efficient operator* of the business upon which a potential purchaser would be likely to base an offer. Where trading information is provided by the actual operator the valuer should not assume that it represents the fair maintainable trade. Just as a valuer of investment *property* should test whether a passing rent is in-line with current open *market rent*, a *trade related property* valuer should test by reference to comparables whether the present trade represents a reasonably maintainable trade in current market conditions

4.7 When assessing future trading potential, the valuer should exclude any turnover and profit that is attributable solely to the personal skill, expertise, reputation and/or brand name of a licensor, the existing owner or management. However, in contrast, the valuer should include any additional trading potential which might be realised under the management of a *reasonably efficient operator* taking over the existing business at the *date of valuation*.

4.8 When valuing *properties* by reference to trading potential, the valuer will need to compare trading profitability with similar types and styles of operation. Therefore a proper understanding of the profit potential of those *property* types, and how they compare to one another, is essential.

4.9 The valuer should endeavour to establish the accuracy and reliability of trading information provided for the purpose of the *valuation*. If any doubt of its accuracy exists, or of the underlying *assumptions* supplied, the valuer should recommend verification.

4.10 For many trading entities the vehicle for a transfer of the going concern business will be the sale of a freehold or leasehold interest in *property*. Such transactional evidence can be used as comparable evidence in the *valuation of trade related properties*, so long as the valuer is in a position to exclude the value of the component parts of the transaction that are not relevant to the *valuation* of a *trade related property*, for example stock, consumables, cash and intangible assets.

4.11 A secondary basis of comparison may be by reference to physical factors, for example, when comparing one hotel with another using a value per bedroom approach. However, when using such a method it is essential that the basis used for comparison is truly relevant, as regards style, location, trading circumstances, and so on.

4.12 New competition can have a dramatic effect on profitability, and hence value. The valuer should be aware of the impact of current, and expected future, levels of competition and, if a significant change from existing levels is anticipated, should clearly identify this in the *report* and comment on the general impact it might have on profitability and value.

4.13 Outside influences, such as the construction of a new road or changes in relevant legislation, can also result in a very substantial effect on the value of *property* valued with regard to its trading potential.

4.14 Particular care must be taken, where the *valuation* is for the purposes of *financial statements*, to ensure that other items in the *financial statements* are not already included in the valuation.

Regulatory licences, consents, certificates, permits and approvals

4.15 When *properties* are sold as fully-equipped operational entities, the purchaser will normally need to renew licences or other statutory consents and take over the benefit of existing approvals, certificates and permits. For t*rade related property valuations* it is normally assumed that all licences and permits will be transferred or granted on the date of transfer of the *property* interest, so where the valuer is making any different assumption it should be clearly stated as a *special assumption*.

4.16 The valuer should, where possible, inspect the licences, approvals, consents, permits and certificates relating to the *property*. Where this is not possible the *assumptions* made should be identified in the *report*, together with a recommendation

Allocation (or apportionment) of valuations or transaction prices to their components

5.1 Where a *valuation* of a *trade related property* has been provided as a fully-equipped operational entity the valuer may be asked to provide an apportionment of the *valuation* between the elements listed in paragraph 2.3.

5.2 Where a transaction includes *trade related property*, or there has beensome other form of business transfer, the valuer may be requested to allocate a transaction price into any or all of the following components:

- the elements listed in paragraph 2.3;
- inventory, being the *trading stock* and consumables;
- intangible *assets* and liabilities. It is possible that several types of intangible *assets* will have to be separately identified and these could include:
  - marketing-related;
  - customer-related;
  - artistic-related;
  - contract-based; and
  - technology-based.

In some cases these allocations may also be required for the purpose of secured lending or *property* taxation.

5.3 Before considering making such apportionments it is essential that the valuer has detailed knowledge of the market in the particular type of trade and is able to distinguish between the value of *transferable goodwill* and *personal goodwill*.

6. Going concern or business valuations

6.1 This guidance note does not apply to going concern or business *valuations*.

6.2 In such cases the guidance in IVS GN 6 — *Business valuation* should be considered.

6.3 Members who carry out trade related *valuations* may also be active in the *valuation* and transfer of businesses. These members acquire specialist knowledge, skills and access to relevant data which enable them to undertake such instructions.

6.4 This guidance note does not deal with the *valuation* of intangible *assets* and the valuer will need to liaise with other consultants who have specialist knowledge of analysing the business's financial performance, and valuing intangible.

Purposes for valuation

There are six main reasons why valuations are required, which we detail below:

(i) **Transaction advice (both capital values and annual values)**
When agreeing to buy or sell a hotel it is usual that both parties are keen to know the value of the property, to ensure they do not sell it too cheaply or buy it too expensively. The same is true when a lease is being agreed; if either party does not have competent advice as to the rental value of the property then the rent may either be too high (thereby harming the long-term profitability of the hotel operator), or too low (thereby diminishing the value of the landlord's interest).

(ii) **Secured lending**
Most hotels are purchased, with a bank providing a large proportion of the value of the property to the owner by way of secured debt, much the same way that a residential property is purchased with a mortgage. For a bank to lend the money on the property at competitive lending rates, it will require a valuation to be carried out on the property to assess its value.

(iii) **Property taxation**
In a number of countries, including the UK, property tax is based upon the notional rental value of the property and as such both the government and the owner need to be able to calculate the liability accurately to ensure that neither too much nor too little tax is being collected.

(iv) **Company accounts**
A number of listed companies revalue their hotel assets on a regular basis and these values are reported within their company accounts. Without accurate assessments of the value of the properties, it is likely that the company's share price would not be able to be accurately assessed.

(v) **Internal purposes**
It is sometimes important for companies to review their assets for purely strategic reasons; in these cases, the valuer will be looking to meet the requirements of the client and will not be bound by the usual valuation guidelines unless, of course, the client requires a Market Value.

(vi) **Company taxation**
It is necessary to determine the depreciation required on the asset for accounting purposes. This will ensure that the correct amount is deducted for tax purposes.

Disposal/purchase — transaction advice on capital values

The first category of "valuation purpose" is one of the most commonly required in the hotel world. It is when a valuation is required if an owner wishes to sell a property, or if someone is keen to buy a property. The basis of valuation is Market Value, which is defined in the *Red Book* under Practice Statement 3.2 as follows:

> The estimated amount for which a property should exchange on the date of valuation between a willing buyer and a willing seller in an arm's-length transaction after proper marketing wherein the parties had each acted knowledgeably, prudently and without compulsion.

The *Red Book* also comments on the definition as follows:

> 3.2 The term property is used because the focus of these Standards is the valuation of property. Because these Standards encompass financial reporting, the term Asset may be substituted for general application of the definition. Each element of the definition has its own conceptual framework.
>
> 3.2.1 'The estimated amount . . .'
> Refers to a price expressed in terms of money (normally in the local currency) payable for the property in an arm's-length market transaction. Market Value is measured as the most probable price reasonably obtainable in the market at the date of valuation in keeping with the Market Value definition. It is the best price reasonably obtainable by the seller and the most advantageous price reasonably obtainable by the buyer. This estimate specifically excludes an estimated price inflated or deflated by special terms or circumstances such as atypical financing, sale and leaseback arrangements, special considerations or concessions granted by anyone associated with the sale, or any element of Special Value.
>
> 3.2.2 '. . . a property should exchange . . .'
> Refers to the fact that the value of a property is an estimated amount, rather than a predetermined or actual sale price. It is the price at which the market expects a transaction that meets all other elements of the Market Value definition should be completed on the date of valuation.
>
> 3.2.3 '. . . on the date of valuation . . .'
> Requires that the estimated Market Value is time-specific as of a given date. Because markets and market conditions may change, the estimated value may be incorrect or inappropriate at another time. The valuation amount will reflect the actual market state and circumstances as of the effective valuation date, not as of either a past or future date. The definition also assumes simultaneous exchange and completion of the contract for sale without any variation in price that might otherwise be made.
>
> 3.2.4 '. . . between a willing buyer . . .'
> Refers to one who is motivated, but not compelled to buy. This buyer is neither over-eager nor determined to buy at any price. This buyer is also one who purchases in accordance with the realities of the current market and with current market expectations, rather than on an imaginary or hypothetical market which cannot be demonstrated or anticipated to exist. The assumed buyer would not pay a higher price than the market requires. The present property owner is included among those who constitute 'the market'. A valuer must not make unrealistic Assumptions about market conditions or assume a level of Market Value above that which is reasonably obtainable.
>
> 3.2.5 '. . . a willing seller . . .'
> Is neither an over-eager nor a forced seller prepared to sell at any price, nor one prepared to hold out for a price not considered reasonable in the current market. The willing seller is motivated to sell the

property at market terms for the best price attainable in the (open) market after proper marketing, whatever that price may be. The factual circumstances of the actual property owner are not a part of this consideration because the 'willing seller' is a hypothetical owner.

3.2.6 '... in an arm's-length transaction ...'
Is one between parties who do not have a particular or special relationship (for example, parent and subsidiary companies or landlord and tenant) which may make the price level uncharacteristic of the market or inflated because of an element of Special Value (see IVS 2, paragraph 3.8). The Market Value transaction is presumed to be between unrelated parties each acting independently.

3.2.7 '... after proper marketing ...'
Means that the property would be exposed to the market in the most appropriate manner to effect its disposal at the best price reasonably obtainable in accordance with the Market Value definition. The length of exposure time may vary with market conditions, but must be sufficient to allow the property to be brought to the attention of an adequate number of potential purchasers. The exposure period occurs prior to the valuation date.

3.2.8 '...wherein the parties had each acted knowledgeably, prudently ...'
Presumes that both the willing buyer and the willing seller are reasonably informed about the nature and characteristics of the property, its actual and potential uses, and the state of the market as of the date of valuation. Each is further presumed to act for self-interest with that knowledge and prudently to seek the best price for their respective positions in the transaction. Prudence is assessed by referring to the state of the market at the date of valuation, not with benefit of hindsight at some later date. It is not necessarily imprudent for a seller to sell property in a market with falling prices at a price which is lower than previous market levels. In such cases, as is true for other purchase and sale situations in markets with changing prices, the prudent buyer or seller will act in accordance with the best market information available at the time.

3.2.9 '... and without compulsion.'
Establishes that each party is motivated to undertake the transaction, but neither is forced or unduly coerced to complete it.

3.3 Market Value is understood as the value of a property estimated without regard to costs of sale or purchase, and without offset for any associated taxes.

The *Red Book* provides further commentary to explain the detail behind the valuation.

Commentary

1. The basis of *Market Value* is an internationally recognised definition. It represents the figure that would appear in a hypothetical contract of sale at the *valuation date*. Valuers need to ensure that in all cases the *basis* is set out clearly in both the instructions and the report.

2. *Market Value* ignores any existing mortgage, debenture or other charge over the *property*.

3. In the conceptual framework in IVS quoted above (para 3.2.1) it is clear that any element of *special value* that would be paid by an actual *special purchaser* at the *date of valuation* must be disregarded in an estimate of *Market Value*. *Special value* includes *synergistic value*, also known as *marriage value*.

4. IVS describes *special value* and *synergistic value* as follows:

Special Value can arise where an asset has attributes that make it more attractive to a particular buyer, or to a limited category of buyers, than to the general body of buyers in a market. These attributes can include the physical, geographic, economic or legal characteristics of an asset. Market Value requires the disregard of any element of Special Value because at any given date it is only assumed that there is a willing buyer, not a particular willing buyer. Synergistic Value can be a type of Special Value that specifically arises from the combination of two or more assets to create a new asset that has a higher value than the sum of the individual assets. When Special Value is reported, it should always be clearly distinguished from Market Value.

5. Notwithstanding this general exclusion of *special value* where the price offered by prospective buyers generally in the market would reflect an expectation of a change in the circumstances of the *property* in the future, this element of 'hope value' is reflected in *Market Value*.
Examples of where the hope of additional value being created or obtained in the future may impact on the *Market Value* include:

- the prospect of development where there is no current permission for that development; and
- the prospect of '*synergistic value*' arising from merger with another *property* or interests within the same *property* at a future date.

6. When *Market Value* is applied to *plant & equipment*, the word '*asset*' may be substituted for the word '*property*'. The valuer must also state, in conjunction with the definition, which of the following additional assumptions have been made:

- that the *plant & equipment* has been valued as a whole in its working place; or
- that the *plant & equipment* has been valued for removal from the premises at the expense of the purchaser.

Further information on *plant & equipment valuation*, including typical further *assumptions* that may be appropriate in certain circumstances, can be found in GN 2 and in IVS GN 3 — *Plant & equipment*.

7. Where the *property* includes land which is mineral bearing, or is suitable for use for waste management purposes, it may be necessary to make *assumptions* to reflect either the potential for such uses or, where the land is already in such use, to reflect any potential future uses that may be relevant. Further information on the *valuation* approach in these cases can be found in GN 4.

8. Where the *property* is personal *property* it may be necessary to interpret *Market Value* as it applies to different sectors of the market. Further information on this type of *valuation* can be found in IVSC GN 4 and 5.

As such, it is commonplace for valuations to be carried out as "market value as a fully equipped trading entity" or some other such category meaning that it is being valued as an operational hotel subject to its potential trading profit.

There are some instances when a valuer is asked to value a property subject to specific assumptions, for example, assuming that a proposed redevelopment plan has been completed, or assuming certain trading levels have been reached. These are valuations subject to special assumptions and ideally the valuer will always report the Market Value prior to providing any Market Values subject to special assumptions.

It could be that these specific assumptions lead the valuer to believe that the resulting value will not reflect market value subject to special assumptions and instead will only reflect the "worth" to that specific owner (it is a worth calculation if the assumptions are not those that would be reasonable for the reasonably efficient operator to adopt). In this instance, the calculation must be clearly labelled as an estimate of worth and the valuer must bring this to the client's attention.

What buyers and sellers want from a valuation?

As simple as it seems, what both buyers and sellers really require is information as to the value of the property so they know how the agreed sales price relates to the value of the property.

In most instances, the purchaser wants to be comforted that they are not overpaying for the property and the seller wants to know that they have not agreed to sell it too cheaply.

Capital value — secured lending purposes

In the majority of property transactions, the finance for the purchase of the property is only partly funded with equity from the purchaser; the balance is usually funded through a secured loan from one of the many potential banking institutions. This is an incredibly simplistic summary, as there are many different sources of debt requiring various levels of interest based upon the level of security they have over the property. However, for the purposes of this type of valuation it is explanation enough.

The theory behind the requirement for secured lending goes straight to the heart of the theory of capitalism. To ensure that the supply of money available in the market does not keep expanding, a certain proportion of all loans have to be placed on deposit. This is required because if a bank could lend 100% of the money that was placed on deposit with it there could be a never-ending spiral in the supply of money. If the debt is unsecured, then 100% of the loan needs to be placed on deposit. As the security level of the loan increases, the amount needed to be placed on deposit decreases. A loan secured on a property may require 50% of the loan to be placed on deposit, whereas a loan on a property that has an appropriate valuation showing that the lending amount is covered by the value of the asset will require a substantially lower amount of the loan to be placed in the vault. As such, a lender who has the benefit of a formal valuation will be able to offer more competitive rates to their customers as the loan is effectively less expensive to them, and the savings can be passed to their customer.

The debt provider will usually be supplying the majority of the money for the transaction (loan to value ratios for hotels differ depending upon affordability and serviceability). As such, the valuer has a significant duty of care to the lending institute and it is usual to find that the bank becomes the valuers' client, no matter who first agreed the terms for the valuation or indeed pays the valuation fee.

The *Red Book* is quite prescriptive in what a valuer has to do when undertaking a valuation for secured lending purposes, as detailed below:

2.1 Before accepting instructions, the *member* must disclose to the lender any anticipated, current or recent fee-earning involvement with the *property* to be valued, with the borrower or prospective borrower, or with any other party connected with a transaction for which the lending is required. In this context 'recent involvement' would normally be anything within the past two years, but under certain circumstances could be longer. For further guidance see appendix 1.1 — *Conflicts of interest*.

2.2 The *terms of engagement* (PS 2.1) should reflect the particular requirements of the lender, if these are consistent with these standards. To avoid any doubt, any *special assumptions* (PS 2.3) that the valuer anticipates may be necessary must be agreed and recorded in the *terms of engagement*. The *special assumptions* will vary depending on the type of proposition and the nature of the *property*.

2.3   The following are the most common examples of security where a valuer's advice is likely to be sought:

- *property* which is, is to be, or has been, owner-occupied;
- *property* which is, or is to be, held as an investment;
- *property* which is equipped as an operational entity and which is generally valued according to trading potential; and
- *property* which is, or is intended to be, the subject of development or refurbishment.

Specific matters which should be considered and included in *valuation* and *appraisal* reports for these different situations are included under Reporting, below.

Where the *valuation* is of a single residential unit that is being acquired with a view to letting as an investment (commonly known as 'buy to let'), the lender's instructions will usually be in accordance with UKPS 3.2 — *Residential property mortgages*, and UK appendix 3.2 — *RICS mortgage valuation specification*.

2.4   The valuer should enquire if there has been a recent transaction, or a provisionally agreed price, on any of the *properties* to be valued. If such information is revealed, enquiries that are practicable as to the circumstances of that transaction should be made. For example, the extent to which the *property* was marketed, the effect of any incentives, and whether the price realised was the best price obtainable.

3. Basis of value

3.1   *Market Value* is the appropriate *basis of value* that should be used for all valuations or appraisals undertaken for secured lending. Any *special assumptions* made in arriving at the *Market Value* must be agreed with the lender in advance and referred to in the report. If the valuer is aware of any other circumstances which could affect the price, these must also be drawn to the attention of the lender, and an indication of their effect provided.

3.2   Examples of circumstances that often arise in *valuations* for secured lending where *special assumptions* may be appropriate include:

- the anticipation of planning consent for development at the property;
- the anticipation of a physical change to the *property*, for example, new construction or refurbishment;
- the anticipation of a new letting on given terms, or the settlement of a rent review at a specific rent;
- the existence of a *special purchaser*, which may include the borrower;
- a known constraint which could prevent the *property* being either brought to, or adequately exposed to, the market;
- any other factor which potentially conflicts with the definition of *Market Value* or its underlying assumptions, as set out in the supporting commentary in PS 3.2;
- the anticipation of a new economic or environmental designation.

This list is not exhaustive and the appropriate *special assumptions* will depend on the circumstances under which the *valuation* is requested, and the nature of the *property* to be valued.

3.3   In all cases (except where the *property* to be valued is equipped as an operational entity) any additional value attributable to goodwill, or to fixtures and fittings which are only of value in situ to the present occupier, is to be excluded.

The Purpose of the Valuation and Statutory Guidance

4. Reporting

4.1 Reports for secured lending should include all the matters set out in PS 6. The following is a guide to other matters that would normally be included in a *report* for secured lending, subject to the exact circumstances of the proposed loan, and the detailed requirements of the lender:

- comment on potential and demand for alternative uses, or any foreseeable changes in the current mode or category of occupation;
- disrepair, or whether any deleterious or harmful materials have been noted;
- contamination or environmental hazards noted;
- comment on past, current and future trends in the *property* market in the locality and/or demand for the category of *property*;
- comment on both the current marketability of the interest and whether this is likely to be sustainable over the life of the loan;
- the *valuation* methodology adopted;
- details of any significant comparable transactions relied upon, and their relevance to the *valuation*;
- comment on the suitability of the *property* as security for mortgage purposes, bearing in mind the length and terms of the loan contemplated;
- if appropriate, a statement drawing attention to any other matter revealed during normal *valuation* enquiries which might have a material affect on the value currently reported;
- comment on any environmental or economic designation.

4.2 Where a recent transaction, or a provisionally agreed price has been disclosed, the valuer is to indicate in the *report* the extent to which that information has been accepted as evidence of *Market Value*. Where the enquiry made under paragraph 2.4 does not reveal any information, the valuer will make a statement to that effect in the report, accompanied by a request that if such information comes to light before the loan is finalised, the matter must be referred back to the valuer for further consideration.

4.3 The following paragraphs deal with the most common lending situations, and indicate matters which should normally be included in an *appraisal* for loan security, in addition to the minimum requirements set out in PS 6 and in the Reporting section, above, together with examples of *special assumptions* that may typically be appropriate.

The *Red Book* specifies additional compulsory requirements for secured lending valuations on investments as detailed below:

- summary of occupational leases, indicating whether the leases have been read or not, and the source of any information relied on;
- current rental income, and comparison with current market rental value. Where the *property* comprises a number of different units, which can be let individually, separate information should be provided on each;
- comment on the market's view of the quality, suitability and strength of the tenant's covenant;
- comment on sustainability of income over the life of the loan, with particular reference to lease breaks or determinations, and anticipated market trends;
- comment on any potential for redevelopment or refurbishment at the end of the occupational lease(s).

The *Red Book* further specifies that properties that are fully equipped as a trading entity and valued with regard to trading potential should reflect the following:

Typical special assumptions:
- *assumptions* made of the trading performance.

Closure of the business could have a significant impact on the *Market Value*. The valuer should therefore *report* the effect that closure of the business would have on the *Market Value*, referring to any appropriate *special assumptions* listed below:

- the business had been closed;
- the inventory had been depleted or removed;
- licences had been lost or had been breached;
- accounts or records of trade would not be available to a prospective purchaser.

Where a valuation is required on the Special Assumption that the work had been completed the value reported should be the current value rather than a projection, or valuation forecast, of the likely value at the end of the development period.

Another important item of statutory guidance for valuers is the European Mortgage Federation paper on mortgage lending value, which relates mainly to valuations carried out on the European mainland. The detail is provided below but, in essence, the regulations suggest that instead of providing a market value (which is a snapshot at one specific moment in time) the valuer should provide a value that will be valid over a longer time frame as it will be robust enough to take into account market fluctuations.

The *Red Book* makes it clear that the paper is neither mandatory nor approved guidance, but a valuer will need to adopt the standard if required to do so by the lending institution.

> Mortgage Lending Value may be used by the financial services industry in the activity of lending secured by real estate. The Mortgage Lending Value provides a long-term sustainable value limit, which guides internal banking decisions in the credit decision process (e.g. loan-to-value, amortization structure, loan duration) or in risk management.
>
> Mortgage Lending Value facilitates the assessment of whether a mortgaged property provides sufficient collateral to secure a loan over a long period. Given that Mortgage Lending Value is intended to estimate property value for a long period of time, it cannot be grouped together with other valuation approaches used to estimate Market Value on a fixed date.
>
> Additionally, Mortgage Lending Value can be used as a risk management instrument in a number of ways in the context of:
>
> - capital requirements for credit institutions as detailed in Basle I and II;
> - funding of mortgage loans through covered bonds secured by real estate as the cover assets;
> - the development of capital market products converting real estate and real estate collateral into tradable assets (e.g. mortgage backed securities).
>
> The concept of Mortgage Lending Value is defined in detail by legislation, Directives and additional country specific regulations.
>
> Mortgage Lending Value shall mean the value of the property as determined by a Valuer making a prudent assessment of the future marketability of the property by taking into account the long-term sustainable aspects of the property, the normal and local market conditions, as well as the current use and alternative possible uses of the property. Speculative elements should not be taken into account in the assessment of Mortgage Lending Value. Mortgage Lending Value should be documented in a clear and transparent way.

All internationally recognised valuation methods also apply to the Mortgage Lending Value, subject to the type of property and the market specificities (historic, legal etc.) where the property is located. These are:

- comparison method;
- income method;
- depreciated replacement cost method.

Regarding the technical transposition of the definition mentioned above, the long-term validity of Mortgage Lending Value requires compliance with a certain number of steps aimed at eliminating short-term market volatility or temporary market trends. The Valuer must address the following key issues when determining the Mortgage Lending Value of a property:

- The future marketability and saleability of the property has to be assessed carefully and prudently. The underlying time perspective goes beyond the short-term market and covers a long-term period.
- As a principle, the long-term sustainable aspects of the property such as the quality of the location, construction and allocation of areas must be taken into account.
- As far as the sustainable yield to be applied is concerned, the rental income must be calculated based on past and current long-term market trends. Any uncertain elements of possible future yield increases should not be taken into account.
- The application of capitalization rates is also based on long-term market trends and excludes all short-term expectations regarding the return on investment.
- The Valuer must apply minimum depreciation rates for administration costs and capitalization of rents.
- If the Mortgage Lending Value is derived using comparison values or depreciated replacement costs, the sustainability of the comparative values needs to be taken into account through the application of appropriate discounts where necessary.
- The Mortgage Lending Value is generally based on the current use of the property. The Mortgage Lending Value shall only be calculated on the basis of a better alternative use, under certain circumstances, i.e. if there is a proven intention to renovate or change the use of the property.
- Further requirements, for example with respect to compliance with national standards, transparency, content and comprehensibility of the valuation, complement the legal framework for the calculation of Mortgage Lending Value.

Major banks and other lenders are subject to regulations that limit the total amount they can lend as a proportion of their assets. This is known as the "solvency ratio". In the international context, the Basle Committee on Banking Supervision has issued an Accord, Basle II, which sets out agreed minimum solvency ratios to be maintained by lending institutions and how those ratios are to be calculated. These are enforced through national laws and, in the case of the European Union, in accordance with EU Directives. The value of assets over which the lender holds security is used in calculating the solvency ratio.

The Basle II Accord provides that one of two valuation approaches, to which different risk criteria apply, may be adopted to assess the value of security represented by commercial real estate:

- The market value of the secured assets; or

- Mortgage lending value.

Mortgage lending value (MLV) is a long-term risk assessment technique, rather than a valuation. The detailed application of MLV may vary from country to country and, before accepting an instruction to calculate MLV, members should ensure that they are familiar with any relevant requirements in that specific jurisdiction, including any restrictions on who may undertake this work. For example, MLV is particularly common in Germany, where banks require individuals calculating MLV to be certified in accordance with a national scheme and to apply a methodology set out under federal regulations.

What banks need from a secured lending valuation report?

- To learn the value of the hotel — this may seem obvious, but lenders need to know the value of the asset they are lending against to be aware of their "solvency ratio", rather than to apply a slightly outdated loan-to-value ratio as part of their lending criteria.
- To know what the asset is likely to generate if the property needs to be sold quickly, hence they request valuations subject to special assumptions (see Chapter 12).
- To have an independent expert assess the hotel and the business and provide them with a commentary on the strengths and weaknesses of the hotel, and set it in an appropriate context for the market in which it trades.
- To have an independent expert review the trading projections of the borrower, which will help the bank determine the serviceability of the debt.
- To have someone to sue if everything goes wrong.

It is not easy to know what the bank requires from the valuer; it could be any combination of the above. To ensure that the bank's requirements are met, it is good practice to spend some time talking to the lender about their specific requirements, and then discussing the results with the bank when the work is done. That way, it will be possible to explain the key points of the report and valuation, highlighting any areas of concern, any potential risks for the future, and any opportunities for their client.

Capital value — company accounts purposes

Once again, the *Red Book* provides a good basis for how to carry out a valuation of a hotel (as a non-specialised property) for inclusion within reported company accounts. Practice Statement 1.1 states that:

> Valuations for inclusion in *financial statements* prepared in accordance with UK Generally Accepted Accounting Principles (UKGAAP) shall be on the basis of either:
>
> (a) *Properties* other than *specialised properties*:
> Existing Use Value (EUV), as defined in UKPS 1.3, for *properties* that are owner occupied for the purposes of the entity's business; or
> *Market Value* (MV), as defined in PS 3.2, for *property* that is either surplus to an entity's requirements or held as an investment;
>
> (b) For *specialised properties*:
>
> Depreciated replacement cost.
>
> *Valuations* for this purpose are *regulated purpose valuations* (see UKPS 5.1) and the various disclosure requirements within PS 5.2 to 5.4 will apply.

The Purpose of the Valuation and Statutory Guidance

Valuations based on Existing use value (EUV) shall adopt the definition settled by RICS. Existing Use Value is to be used only for valuing *property* that is owner-occupied by a business, or other entity, for inclusion in *financial statements*.

Definition
'The estimated amount for which a *property* should exchange on the *date of valuation* between a willing buyer and a willing seller in an arm's-length transaction, after proper marketing wherein the parties had acted knowledgeably, prudently and without compulsion, assuming that the buyer is granted vacant possession of all parts of the *property* required by the business and disregarding potential alternative uses and any other characteristics of the *property* that would cause its *Market Value* to differ from that needed to replace the remaining service potential at least cost.'

Commentary
1. The definition of EUV is the MV definition with one additional *assumption* and a further requirement to disregard certain matters. The definition must be applied in accordance with the conceptual framework of MV at PS 3.2, together with the following supplementary commentary:

2. '... *the buyer is granted vacant possession* ...'
2.1 The *assumption* that vacant possession would be provided on acquisition of all parts of the *property* occupied by the business does not imply that the *property* would be empty, but simply that physical and legal possession would pass on completion. Any parts of the *property* occupied by third parties should be valued subject to those occupations. *Properties* occupied by employees, ex-employees, or their dependants, should be valued with regard to the circumstances of their occupation, including any statutory protection.

2.2 This *assumption* also means that it is not appropriate to reflect any possible increase in value due to special investment or financial transactions (such as sale and leaseback) which would leave the owner with a different interest from the one which is to be valued. In particular the covenant of the owner-occupier must be ignored.

3. '... *of all parts of the property required by the business* ...'
3.1 If parts of the *property* are unused, and surplus to the operational requirements of the business, their treatment will depend on whether they could be sold or leased separately at the valuation date. If they could be occupied separately then they should be allocated to a separate category as surplus *property*, and valued on the basis of MV. If separate occupation is not possible any surplus parts would have no more than a nominal Existing Use Value, as they would contribute nothing to the service potential of the *property*, and would not feature in a replacement at least cost.

4. '... *disregarding potential alternative uses* ...'
4.1 Existing use, in the context of EUV, means that the valuer should disregard uses that would drive the value above that needed to replace the service potential of the *property*. An entity seeking to replace the service potential of the asset at least cost will not buy a *property* if its value has been inflated by bids from other potential occupiers for whom the *property* has greater value because of alternative uses or development potential which are irrelevant to its own requirements.

4.2 The valuer must ignore any element of hope value for alternative uses which could prove more valuable, though it would be appropriate to take into account any value attributable to the possibility of extensions or further buildings on undeveloped land, or redevelopment or refurbishment of existing buildings, providing that these would be required and occupied by the entity, and that such construction could be undertaken without major interruption to the current operation.

Valuation of Hotels for Investors

5. *'... disregarding ... any other characteristics of the property that would cause its Market Value to differ from that needed to replace the remaining service potential at least cost.'*

5.1 There are circumstances where it may be appropriate for the valuer to ignore factors that would adversely affect the *Market Value*, but would not be characteristic of a replacement. Examples include:

- where an occupier is operating with a personal planning consent that could restrict the market in the event of the owner vacating;
- where the occupier holds the property under a lease and there are lease covenants that impose constraints on assignment or alternative uses;
- where a *property* is known to be contaminated, but the continued occupation for the existing use is not inhibited or adversely affected, provided there is no current duty to remedy such contamination during the continued occupation;
- where an industrial complex is overdeveloped, and the extra buildings have either a limited *Market Value*, or detract from the *Market Value*, but would need to be replaced to fulfil the service potential to the business;
- where the existing buildings are old, and so have a limited *Market Value*, but would have a higher replacement cost to the business;
- where the *property* is in an unusual location, or oversized for its location, with the result that it would have a very low *Market Value*, but where the cost of replacing the service potential would be significantly greater;
- where the market is composed solely of buy-to-let investors, but the valuer believes that the replacement cost (the price agreed between a willing vendor and willing owner-occupier purchaser) would be higher.

5.2 Any value attributable to goodwill should normally be ignored, with the exception of *trade related property* (see GN 1), where the element of goodwill that is reflected in the trading potential (that which is inseparable from the interest in the *property*) should be included in the EUV.

5.3 The fact that a large *property* may be in single occupation does not necessarily mean that it has to be valued on the assumption that only bids from other potential occupiers for the whole can be taken into account. If the *property* is one where a higher value would be generated by the potential to divide it into smaller units for the existing use, this should be reflected in the *valuation*.

6. The underlying accounting concepts behind EUV are discussed in UK appendix 1.1.

7. Whilst the definition of EUV has changed from that published in previous editions of the *Red Book*, the underlying principles have not. The previous definition was based on the OMV definition, which has now been removed from these standards, but the new definition uses the MV definition, and the additional provisions have been reworded to define more precisely the requirements of the accounting standard. An EUV provided under the old or the new definition should produce exactly the same result.

8. Many market *valuations* are based on the existing planning use of the *property* as this represents the use which generates the highest value. Such *valuations* have sometimes been described as existing use *valuations*. However, this is incorrect and these valuations should properly be expressed as *Market Values*. It is emphasised that existing use value as defined is for use only when valuing, for inclusion in *financial statements*, *property* that is occupied by the owners of the interest being valued for the purpose of their business.

9. Further guidance on the interpretation of EUV is contained in Valuation Information Paper 1 — *Valuation of Owner-Occupied Property for financial statements*.

The Purpose of the Valuation and Statutory Guidance

1.4 Differences between EUV and MV

Where there is a significant difference between the EUV and MV of the same *property* if it were surplus the *member* must provide an opinion on both bases and explain the reasons for this in the report. Any published reference to the *report* must refer to the *valuations* on both bases, except where the difference has no material effect on the aggregate value of the entity's *properties*.

1. Where the *property* is of a type that is commonly traded in the market, with no higher value for an alternative use and with no unusual features that could restrict marketability, there would normally be little material difference between MV and EUV. However, as will be clear from the commentary to UKPS 1.3, there will be cases where the difference between the two bases is material. Where a difference exists, this could be material to an overall assessment of the entity's financial position, and in such cases the valuer must *report* both bases. In the case of a company there is an obligation (under the *Companies Act* 1985, para. 1(2) Schedule 7) for the directors to disclose if the *Market Value* of its *assets* is materially different from the figure that appears in the balance sheet.

2. Where the *valuation* involves a *portfolio of properties*, and the difference between EUV and MV on an individual *property* does not make a material difference to the aggregate value of the *properties*, the valuer simply needs to indicate whether the MV of a particular *property* would be higher or lower than the reported EUV. There is no obligation to provide alternative *valuations* or reasons for the difference.

The *Red Book* goes even further when discussing valuations for inclusion within financial statements, stating:

Valuations for *financial statements* prepared under *International Financial Reporting Standards* (IFRS) shall be in accordance with the IVSC *International Valuation Application* 1 (IVA 1).
Commentary

1. IVA 1 is published by the IVSC as part of the International Valuation Standards. It contains information on the following *International Financial Reporting Standards*:

    - IAS 2 (Properties Held for Sale);
    - IAS 16 (Property, Plant & Equipment);
    - IAS 17 (Leases);
    - IAS 36 (Impairment of Assets);
    - IAS 40 (Investment Property);
    - IAS 41 (Agriculture);
    - IFRS 3 (Business Combinations) and
    - IFRS 5 (Non-Current Assets Held for Sale and Discontinued Operations).

    Appendix 4.1 contains the full text of IVA 1. Appendix 6.3 contains additional guidance on reporting *valuations* under IFRS.

2. *Valuations* should be reported at *Market Value*. Any *assumptions* or qualifications made in applying *Market Value* should be discussed with the entity and disclosed in the report.

3. Although under IVA 1 the valuer will *report* the *Market Value*, the entity is required under IFRS to account for the *asset* at its *fair value*. To assist the entity in this respect and to enable it to make the disclosures required under IAS 16 and IAS 40, the *report* must contain the following information:

- the effective *date of* the (re)*valuation* (see PS 2.1 (g));
- whether the valuer is an external or *internal valuer* (see PS 1.6 and 1.7);
- the methods and significant *assumptions* applied in estimating the *Market Value* (see PS 6.1 (f) and (k), PS 2.4, and PS 6.4 to PS 6.7);
- under IAS 16: The extent to which the values were determined by reference to observable prices in an active market or recent market transactions on arm's length terms or were estimated using other *valuation* techniques (see appendix 6.1(p));
- under IAS 40: The method and significant *assumptions* applied in determining the value of investment *property*, including a statement whether the determination of *fair value* was supported by market evidence or was more heavily based on other factors (which the entity shall disclose) because of the nature of the *property* and lack of comparable market data (see appendix 6.1(q)).

4. Where the *valuation* has been prepared in accordance with IVA 1, a statement to that effect must be made in the report.

In the Department of Trade and Industry report into the collapse of Queens Moat House in 1973 (carried out by Adrian Burn and Patrick Phillips and published in 2004), it concludes:

We recommend that if valuations of this kind are to be incorporated into balance sheets, the bases and assumptions on which they have been calculated should be clearly stated in the accounts. We acknowledge that, with a large number of hotels, some of these assumptions might need to be expressed within certain ranges but, nevertheless, if they are to be used in company accounts and are material, they warrant clearer explanation of how they have been arrived at.

However, the report warns about problems of including annual valuations into company accounts:

The inherent sensitivity of hotel valuations may not be fully understood by users of accounts. In the case of QMH, shareholders may have been deluded into thinking that their investment was protected by the values given to QMH's hotels even if QMH's trading declined, without realising the dependence of hotel valuations on trading and thus their sensitivity, particularly in times of recession. Indeed some of QMH's public circulars claimed that the valuations showed the "strong asset backing" supporting the acquisition, which suggested that both the professionals associated with the circulars and the users of them may have shared a common misunderstanding.

The report concludes as follows:

It is questionable in our minds, therefore, whether any greater purpose is served by introducing hotel valuations into company balance sheets than would be served by attempting to value goodwill in the balance sheet of any other business. After all, a balance sheet is not, or should not purport to be, a statement of the value of the business, but should be merely showing the deployment of the company's investment in its assets.

What shareholders want from a valuation?

A valuation needs to provide the necessary information to enable shareholders to ascertain the asset value of the company underlying their investment.

Capital value and calculations of worth — internal company purposes/strategic advice

There are times when a client requires a "valuation" to be carried out that is specific to their requirements or includes specific assumptions that would not be typical of the market in general. In this instance, the valuer will not actually be carrying out a valuation but will be undertaking an "estimate of worth".

There is no problem with carrying out any number of worth calculations, and the client can request and the valuer can carry out the most unlikely calculations as long as the valuer is clear throughout the report and when discussing the figures with the client that it is a calculation of worth and not a valuation.

Practice Statement 5.13 of the *Red Book* discusses the concept of worth as follows:

> Where the member provides an assessment of Worth, it must not be described as a Valuation, and a statement must be made that the figure is not a Market Value.

> Commentary

> Worth, or Investment Value, is the value of property to a particular owner, investor, or a class of investors, for identified investment objectives. This subjective concept relates specific property to a specified investor, group of investors, or an entity with identifiable investment objectives and/or criteria. In this context investment includes the benefits of owner occupation.

> Where Worth is being reported, it must not be described as a valuation, and a statement must be included that the figure or figures provided do not represent the Market Value of the property.

Ideally, the valuer would also include details of the Market Value of the property so that the client could see the relationship between the market value and the estimate of worth that has been carried out.

What hotel companies want from their worth calculations?

The requirements for their calculations are likely to be as varied as the special assumptions that are requested, but normally the company is keen to know the potential impact on the asset valuation of certain courses of action.

Rental levels — transactional advice

Although less common than purchases, many transactions now include a hotelier taking a lease of a hotel and, as such, a valuer will be called upon to determine the value of property on an annual basis, which is known as the Rental Value of the property.

The Valuation & Appraisal Manual (*Red Book*) provides the statutory framework in which the valuer will work and states the following:

> Valuations based on Market Rent (MR) shall adopt the definition settled by the International Valuation Standards Committee.

The estimated amount for which a property, or space within a property, should lease (let) on the date of valuation between a willing lessor and a willing lessee on appropriate lease terms in an arm's-length transaction after proper marketing wherein the parties had acted knowledgeably, prudently and without compulsion. Whenever Market Rent is provided the 'appropriate lease terms' which it reflects should also be stated

The *Red Book* also provides the following commentary:

1. The definition of *market rent* is the *Market Value* (MV) definition modified by the substitution of a willing lessor and willing lessee for a willing buyer and willing seller, and an additional *assumption* that the letting will be on 'appropriate lease terms'. This definition must be applied in accordance with the conceptual framework of MV at PS 3.2, together with the following supplementary commentary:

    1.1 '. . . willing lessor and willing lessee . . .'
    The change in the description of the parties simply reflects the nature of the transaction. The willing lessor is possessed with the same characteristics as the willing seller, and the willing lessee with the same characteristics as the willing buyer, save that the word 'price' in the interpretive commentary to MV should be changed to 'rent', the word 'sell' changed to 'let' and the word 'buy' changed to 'lease'.

    1.2 '. . . appropriate lease terms . . .'
    MR will vary significantly according to the terms of the assumed lease contract. The appropriate lease terms will normally reflect current practice in the market in which the *property* is situated, although for certain purposes unusual terms may need to be stipulated. Matters such as the duration of the lease, the frequency of rent reviews, and the responsibilities of the parties for maintenance and outgoings, will all impact on MR. In certain States, statutory factors may either restrict the terms that may be agreed, or influence the impact of terms in the contract. These need to be taken into account where appropriate. Valuers must therefore take care to set out clearly the principal lease terms that are assumed when providing MR.

    If it is the market norm for lettings to include a payment or concession by one party to the other as an incentive to enter into a lease, and this is reflected in the general level of rents agreed, the MR should also be expressed on this basis. The nature of the incentive assumed must be stated by the valuer, along with the assumed lease terms.

    Market rent will normally be used to indicate the amount for which a vacant *property* may be let, or for which a let *property* may re-let when the existing lease terminates. MR is not a suitable basis for settling the amount of rent payable under a rent review provision in a lease, where the actual definitions and *assumptions* have to be used.

Annual valuation — for taxation purposes

The RICS does not provide any formal guidance on how to undertake a valuation for rating purposes, although the Valuation Office (VO) has provided an outline paper on how to carry out hotel valuations for this purpose.

The VO provides the following advice for calculating the rating assessment of hotels:

> Rental evidence should form the basis of valuation. However, due to the paucity of unimpeachable evidence it will usually be necessary for most offices to have regard to all national rents.

Where rental evidence exists, it should be devalued by reference to receipts, as a percentage of Gross Takings All Sources (GTAS) excluding VAT, and in conjunction with an examination of the differential percentages of the main income streams. This analysis, based on a shortened profits method, will form the primary basis of comparison.

A secondary comparison can be made by analysing in terms of gross or net Double Bed Units (DBUs). Both will reflect the occupancy rate of the letting accommodation. The value per gross DBU will also reflect the income earned from restaurant(s), bar(s) and all other ancillary areas and facilities from both residents and non-residents. A more valid analysis uses the net basis after stripping out an amount from the adjusted rent in respect of the restaurant(s), bar(s) etc. Both of these methods of comparison need to be treated with great care.

Shortened R&E (Receipts and Expenditure) Method of Valuation

Where a property is not let or the rent is not reliable, a shortened R&E method should be adopted having regard to the analysis derived from rented properties. The fair maintainable receipts should be ascertained by looking at three consecutive years annual receipts (if available), with the final financial year ending closest to the AVD (whether before or after). Each year's receipts should be adjusted to reflect changes in the economy, price structures etc as well as physical factors affecting the property or its environment.

Receipts and Expenditure (R&E) Method of Valuation

Where full accounts are provided an R&E valuation should be undertaken, and the resultant rental value expressed in terms of a percentage of GTAS and, when possible, a percentage of the main income streams. This analysis of values arrived at by full accounts can be used to support valuations using a shortened method.

Valuers should however be mindful that the valuation must reflect the value of the hereditament and not the business of the actual occupier and should therefore exercise caution before adopting an RV derived from this approach where it falls outside the expected percentage range for the particular type of property.

Owing to the reluctance of hotel operators to divulge accounts and profitability details to the Valuation Office it is not possible to adopt an R&E valuation as the primary valuation method for valuing hotels for rating.

The paper then outlines the following general valuation considerations:

When analysing rental evidence, and comparing or preparing valuations, regard must be had to the income streams and profitability of the revenue sources in a hotel.

The requested information for hotels ask for a division of receipts between:

a) Intoxicating liquor
b) Food — excluding wines and liqueurs
c) Accommodation — excluding meals
d) Other receipts.

Where accounts are available gross profit levels, which are likely to be different for each function, can be measured against performance norms. As precise apportionment of wages and other overheads between each is not possible it is traditional for the accounts to relate to the whole business carried on at the property and the tenants bid will usually be a global percentage reflecting this.
By looking at the trade split, as evidenced by the proportion of income generated by each source, Valuers will however be able to reflect more accurately the differing net profit margins when determining and comparing a single percentage.

Letting accommodation will normally show the highest net profit. The net profit on food will depend on the type of hotel and nature of the catering operation. Normally where it is run as a good class restaurant and many covers are "non-residential" it would be expected to fall between the rates for the wet trade and accommodation. Where the covers are limited to guests, especially for silver service, or alternatively where the restaurant has no pretension, the net profit from catering is likely to be the lowest.

As well as looking at the trade split each income stream can be compared. Accommodation receipts can be analysed per DBU to give a net income per room and catering receipts can be expressed in terms of covers/floor area. There will always be an area of cross-subsidy (eg guests can use leisure facilities which non-residents subscribe to by way of membership fees and/or admission charges) and great care must be taken to reflect these when comparing one hotel with another.

Total receipts can also be analysed per DBU overall to compare the turnover of similar types of hotels. The results must be treated with caution as the analysed figure will reflect all the various facilities available. It will, however, give an indication of whether the hotel is trading significantly above or below the norm for that particular type, although it will not provide any indication of running costs or profitability.

Additional Considerations

The value of a hotel depends on its potential profitability, and where accounts are available for well run hotels these will provide the best evidence. In the absence of full accounts, when making a valuation judgement it will be necessary to consider all factors which could influence either the turnover or operating costs of the hotel in question, or of any comparables. The following list of points for consideration is not exclusive:

Accommodation

It is not sufficient merely to take account of the total accommodation and the following matters should be considered:

whether the number and type of single and double letting bedrooms are in balance with the needs of the locality and the class of trade;

- whether there are sufficient bathrooms (where these are not en-suite with the bedrooms);
- the size, quality and layout of the public rooms;
- whether the kitchens are up to standard and convenient for economical working;
- the adequacy and efficiency of the hot water, heating and air conditioning systems (if any);
- whether there are adequate lifts;
- whether there is sufficient off-street parking;
- whether any of the accommodation is surplus to present day needs.

Location

This is determined by the type of trade envisaged by the hotel operator, or conversely, the location will determine the potential trade of the hotel. Convenience to transport facilities or town centres will encourage tourist and commercial hotels; proximity to airports will lead to the establishment of hotels for executives, and nearness to motorways and other major roads will promote motels and lodges. At seaside holiday resorts the distance from the sea and views from the bedrooms will have a considerable bearing on the popularity of the hotel.

Class of Trade

The "star rating" of the hotel (usually AA) together with any "appointment" by the AA, RAC, or Egon Ronay etc.

Liquor Licence

The nature of the licence (if any).

Catering and Functions

Facilities for catering and restaurant business for non-residents.

Conference Facilities

The size, quality and adaptability of rooms to accommodate business conferences which can be an important adjunct to many of the larger hotels.

Leisure Facilities

Many hotels have leisure facilities such as a swimming pool, health club, gym, squash court and/or golf course. They may be exclusive to residents or open to non-residents, often via a club or membership scheme.

Repairs and Maintenance

Under the definition of Rateable Value these are the responsibility of the hypothetical tenant. The actual condition of the hereditament is therefore important, as are the likely future costs of repairing and maintaining the building and grounds. This is likely to relate closely to the age, construction and design of the property.

Operating Costs

These will depend greatly on the age, layout and services for each individual hotel. An old, sprawling hotel on different levels is likely to be more expensive to light, heat, clean and service than a modern purpose built hotel with a modular layout. Additionally, the operating costs will depend on the level of hospitality and personal service provided; country house hotels will, for example, tend to provide high service levels at correspondingly high cost. The type of hotel will also affect the amount of tenants capital which would be envisaged to be tied up in the property and whether it is likely to be occupied by an hotel group who would have to additionally cover a proportion of head office expenses.

Toning of Receipts

Where receipts information for the three financial years ending closest to the valuation date is unavailable, for example in the case of a recently completed property, or if there has been a change of occupier, or following a MCC (material change of circumstance), it may be necessary to adjust later (or earlier) receipts to the relevant valuation date. Evidence to support the level of adjustment should be drawn from investigation of receipts for similar classes of hotels within the local or regional area. Trade patterns are likely to vary considerably between different types of hotel, and caution should be exercised in applying adjustments gleaned from, for example, modern, purpose built, commercial hotels to older, historic, mixed business/tourist properties.

Valuations for depreciation in company accounts

A company will be able to write off an element of the value of its assets over their expected lifecycle to ensure that it is able to calculate its company tax liability appropriately.

This is a specialist area and can vary from tax jurisdiction to tax jurisdiction; as such, it is essential that the valuer is aware of the exact taxation guidelines that are in place in the location that they are working. For UK properties, guidance for this is found in three key places, the *Red Book*, the British Association of Hotel Accountants (BAHA) and the Financial Reporting Manual (HM Customs).

The *Red Book* states the following:

Where the *member* is required to advise on an apportionment of a *valuation* for depreciation purposes, or to advise on the remaining useful economic life of the asset, the apportionment should be undertaken in accordance with the principles set out at UK appendix 1.4.
Commentary
1. Where land and buildings are occupied by an entity in the normal course of its business the value of those *assets*, as shown in the accounts, may be adjusted to reflect depreciation of those assets over time.

2. The accounting principles for depreciation can be found in FRS 15. These provide that depreciation is applied over the future useful economic life of the *asset* to the entity.

3. As depreciation is normally only applied to *property* occupied by the entity the apportionment will be of EUV, or *depreciated replacement cost*-based *valuations*. Depreciation should not be applied to property valued on the basis of MV, except as provided by SSAP19 in relation to certain investment *properties*.

4. As, in reality, land and buildings are usually inseparable, the apportionment should be reported as being hypothetical and for accounting purposes only.

The *Red Book* goes on to say:

The Accounting Standards Board's IFRS 15 requires entities that revalue their tangible fixed *assets* to carry those *assets* in *financial statements* at current value. Current value is determined using the value to the business model set out in the FRS at appendix IV, paragraph 19. This can be portrayed diagrammatically, as shown below:

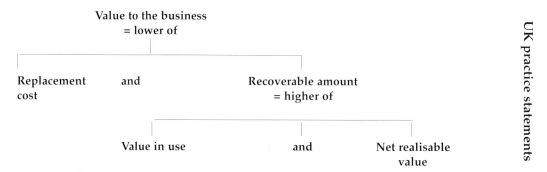

- Replacement cost is the cost of purchasing, at the least cost, the remaining service potential of the *asset* at the balance sheet date. It is an entry value;
- Value in use is the present value of the future cash flows obtainable as a result of an *asset*'s continued use, including those resulting from its ultimate disposal;
- Net realisable value is the amount for which an *asset* could be disposed, less any direct selling costs. It is an exit value.

2.2 Valuers should apply the concept of replacement cost to land and buildings on the following bases:

- *Market Value* (MV) for non-specialised *properties*, but existing use value (EUV) for *properties* which are owner-occupied for the purposes of the business;
- *Depreciated replacement cost* for *specialised properties*, subject to adequate potential profitability.

2.3 The 'value to the business' model prescribes that the value of tangible fixed *assets* in the accounts must be set at a level that is sufficient to reflect the cost in the market of replacing their service potential. This is also referred to as the 'deprival value' of the *asset* where, if the organisation had been deprived of a particular *asset*, the price which it, or any other potential owner occupier for the same use, would pay in the market to replace it to enable the organisation's operations to continue.

2.4 In considering the concept of deprival value in relation to EUV, the actual circumstances of the owner-occupier should not be taken into account as this would be an assessment of *worth*. There is also a risk that the actual owner-occupier could be vested with the characteristics of a purchaser with a special interest, whose bid has to be ignored under the definition of EUV. To avoid reflecting any additional bid that may be made by the actual owner-occupiers because of their particular circumstances, valuers may find it helpful to consider the bid which would be made by a hypothetical purchaser to occupy the *property* for the same use and in a similar manner to the actual occupier. Alternative use values incompatible with the use of the asset in the business have no relevance in the accounts of the company. However, an alternative use that increases the value of a *property* owned and occupied by the entity to a level above that needed to fulfil the service potential may be relevant to an overall *appraisal* of the company's situation, and should be disclosed in the *directors' report* (see paragraph 2, Schedule 7, Companies Act 1985).

2.5 While the value to business model assists valuers in understanding the context in which *valuations* for *financial statements* are required, it should be noted that the use of the word 'value' in the expression 'value in use' does not mean that a valuer is necessarily competent to determine this figure. The term should not be regarded as an alternative *valuation* basis for fixed *assets* and should not be used by valuers when preparing *valuations*. The valuer's role will normally be confined to providing advice on the replacement cost and/or the net realisable value.

2.6 Notwithstanding this caution, members with a particular knowledge of, or skill in, an *asset* class or industry may be competent to assist in the calculation of value in use. Requirements and guidance on the measurement of value in use can be found in FRS 11 — *Impairment of fixed assets and goodwill*.

3. Frequency of valuations

3.1 FRS 15 does not require annual revaluations, although the objective of a revaluation policy is to reflect current values as at the balance sheet date. Full guidance can be found in paragraphs 44–46 and 52 of the FRS but in summary:

- where *properties* are revalued, the requirements of the FRS will be met by a full valuation at intervals of no more than five years and an interim *valuation* in year three;
- if there has been a material change in value, further interim *valuations* should be undertaken in years one, two and four;
- for *portfolios* of *properties*, full *valuations* on a rolling basis may be carried out so that all *properties* are covered over a five-year cycle, subject to interim *valuations* on the remainder of the *portfolio*, where it is likely there has been a material change in value. For this approach to be acceptable, the *properties* must be broadly similar in character, and it must be possible to subdivide the *portfolio* into groups of a broadly similar spread.

3.2 An interim *valuation* may be carried out on a restricted basis, although the FRS makes clear that an *inspection* of the *property* or the locality should still be undertaken 'to the extent that this is regarded professionally necessary'.

3.3 Statement of Standard Accounting Practice 19 (SSAP 19) states that *property* companies holding a substantial proportion of investment *properties* must have an external valuation at least every five years.

Depreciation

Depreciation is defined in IFRS 15 as:

> The measure of the cost, or revalued amount, of the economic benefits of a tangible fixed asset that have been consumed during the period. Consumption includes the wearing out, using up or other reduction in the useful economic life of the tangible fixed asset whether from use, effluxion of time or obsolescence through either changes in technology or demand for goods and services produced by the asset.

FRS 15 requires that depreciation should be allocated on a systematic basis over the future useful economic life of a fixed asset. The depreciation method used should reflect, as fairly as possible, the pattern in which the asset's economic benefits are consumed by the entity. The future useful economic life of an asset is defined in the FRS as the period over which the entity (in whose accounts the asset is carried) expects to derive economic benefit from the asset. All buildings have a limited life due to physical, functional and environmental changes that affect their useful economic life to the business.

As indicated above, the future useful economic life of the tangible fixed asset is defined as the period over which the entity in whose accounts the asset is carried expects to derive economic benefit from that asset. This may be the total physical or economic life of the asset, however, if there is an expectation that the asset will be sold before the end of its physical or economic life, this period will be shorter.

In normal circumstances depreciation is not applicable to freehold or feuhold land. Exceptions to this include land which has a limited life due to depletion, for example, by the extraction of minerals, or which will be subject to a future reduction in value due to other circumstances. One example would arise where the present use is authorized by a planning permission for a limited period, after which it would be necessary to revert to a less valuable use.

Leasehold assets must, by their nature, have a limited life to the lessee, although the unexpired term of a lease may exceed the life of the buildings on the land. Any contractual or statutory rights to review the rent, or determine or extend a lease, must also be considered.

The assessment of depreciation and the remaining useful economic life of the asset are the responsibility of the directors of the company, or their equivalent in other organizations. However Valuers should expect to be consulted on matters relevant to their assessment, such as the degree of obsolescence, condition, market factors, town planning, and so on.

Valued as an operational entity

Where the valuation relates to property valued fully-equipped as an operational entity the valuation figures may need to be apportioned between:

- land;
- buildings;
- fixtures and fittings;
- trading potential.

Paragraph 85 of IFRS 15 suggests that it would not be appropriate to treat the trading potential associated with the property as a separate component of the value of the asset if its value and life were inherently inseparable from that of the property (see also GN1 Trade-related valuations).

Future useful economic life

In order to form an opinion of the future useful economic life of buildings, the Valuer will need to take into account the following matters:

- physical obsolescence — the age, condition and probable costs of future maintenance (assuming prudent and regular maintenance);
- functional obsolescence — suitability for the present use, and the prospect of its continuance or use for some other purpose by the business. In the case of buildings constructed or adapted to meet the requirements of particular uses, including particular industrial processes, the member will need to consult with the directors to ascertain their future plans;
- environmental factors — existing uses must be considered in relation to the present and future characteristics of the surrounding area, local and national planning policies and restrictions likely to be imposed by the planning authority on the continuation of these uses;
- policy on future disposals — the Valuer will need to consult with the entity to ascertain whether there is any intention or policy to dispose of assets before the end of their natural life span.

It is frequently difficult, even impossible, to put a precise life on a building or group of buildings and Valuers may, therefore, have to resort to 'banding'. It should be possible to identify buildings that are unlikely to remain beyond 20 years and other buildings with a life of more than 50 years. In such cases buildings should be noted as having a life of 'not less than 50 years'. Clearly the Valuer's task will be made easier by the use of broad bands and in the majority of cases it is likely these will meet the company's requirements.

Where a property comprises a number of separate buildings, for example, large factory premises, it is suggested that the buildings should be grouped and, wherever possible, a single life allocated to all buildings within each group. Such an approach can be justified by the fact that the life of individual buildings can usually be extended, within reasonable limits, by a higher standard of maintenance or minor improvement. It is normally uneconomic to carry out piecemeal redevelopment.

It would not be appropriate to group buildings if they are used for different industrial processes with different accommodation requirements, or where the client requires that each building must be considered individually.

If consulted on the remaining useful economic life of leaseholds the Valuer must also consider the duration of the lease, any options to determine or extend, the date of the next rent review and whether this is too full or a proportion of rental value.

Summary

There are a wide variety of different purposes for valuation and each have specific professional guidance that needs to be followed to ensure that the valuer provides a professional service to the client.

The guidance that is provided will change depending upon the geographic jurisdiction of the property and the various professional bodies and it is essential that the valuer is clear on what guidance has been followed.

Methodology for Capital Valuations

Introduction

When first learning about valuation it is common practice for students to be introduced to the five methods of valuation, which are outlined below. It is important to note however that all of these rely heavily upon comparable evidence, without which the methodologies would provide nothing more than a mathematical exercise.

It is likely that the hotel valuer will have to work at some time with four of these five methods (comparison, profits, residual and investment) and a thorough understanding of the methodology of each type of valuation is vital.

1. The comparison method

This method requires the valuer to look at comparable sales/lettings and then apply a value to the subject property by adjusting the evidence from similar transactions to meet the criteria of the subject property.

This is most commonly used for houses or flats sold with vacant possession and is usually based on a price per property (two-bedroom, one-reception properties in that block of flats sell for £250,000) or price per square foot (houses in that location sell for £950/psf).

Evidence of sale prices of residential properties are usually freely available and can be analysed quite effectively, albeit that judgment is still required by the valuer to determine the relative advantage and disadvantages of the subject property and the relevant comparables.

In terms of hotels, the comparison method is most commonly used when the property itself, rather than the income it can generate, is the primary driver behind the purchase, for example for lifestyle or some trophy hotels.

2. The investment method

In this method the valuer assumes that the income generated on an annual basis from the property is sustainable in the future (or makes whatever allowances are appropriate) and then applies an appropriate multiplier to this income stream, and deducts transaction costs.

The investment method of valuation is probably the most commonly used method of valuation in commercial property, as the majority of retail, office or industrial transactions occur with the property sold when occupied by a tenant. As such, offices with vacant possession can sometimes be significantly less valuable than as leased-out properties.

It is worth mentioning here that a hotel with vacant possession is not like an office with vacant possession. Hotels are sold as trading entities and "with vacant possession" does not mean vacant or empty, it means without an operating agreement (lease or management contract), but still with all the bookings, staff and other items required to operate the business so the hotel will be generating income immediately. The investment method has become more important in the hotel world since the mid-1990s as more properties are operated on leases management contracts.

3. The profits method

The profits method of valuation requires that the valuer looks at the potential profitability of a property and applies the appropriate multiplier to the net profit or (EBITDA) to assess its value.

The profits method is used when purchasers are attracted to the specific property based upon the profitability of the unit, and a simple comparison with other similar transactions will not provide an accurate enough valuation.

It is the method most commonly used to value properties such as petrol stations as well as hotels. However, it is not used in isolation from the market. The assessment of the potential trading will be based upon the valuer's knowledge of how comparable properties trade, and the multipliers applied to the EBITDA will be based upon comparable evidence. When undertaking the valuation the valuer has to apply significant judgment to complete the valuation.

4. The residual method

The valuer calculates the end value of the completed development, and then deducts all the necessary construction and finance costs, including the developer's profit to calculate the resulting site value.

The residual method of valuation is used to assess site values where there is insufficient comparable evidence of site values, and where developers buy the site based on what the end development will be worth, less the cost of developing the property.

In terms of hotels, many sites are purchased based upon prices that are calculated with reference to the end value of the completed development, because comparable site evidence usually requires a lot of analysis before it is truly useful.

However, this method is still exposed to comparable evidence. The building costs are all market tested, even if the developer has agreed a turnkey development package at a fixed price the valuer will analyse this against "average market costs". In the same way, finance charges, contingency levels and the developer's profits will be linked to normal market parameters.

The end result will then be reviewed, and any alterations that are necessary will be made to ensure the end value is in line with market parameters.

5. The contractors method

In this specialist method of valuation, the valuer looks at the cost of replacing the property (when it is a specialised type of property that very rarely trades on the open market), and assesses its value after reducing the rebuilding cost by an appropriate factor for depreciation and obsolescence.

The contractors method of valuation would be used for public toilets, power stations or other specialist properties where the previous methodologies could not be used.

Hotel valuations

It is worth stressing once again that when carrying out the market valuation to determine the value of the property the valuer is trying to reflect what 'buyers currently active in the market' would pay for a property at that time.

The best way for any valuer to determine what the market will pay is to be thoroughly immersed in the transactional market and to adopt a valuation methodology that best reflects the approach taken by the active purchasers for that type of property at that time.

It can be difficult for a valuer to assess the value of a hotel if they are kept in isolation from deals that are being carried out in the market. The hotel market changes quickly, with new buyers moving in from various sectors and each having their own purchasing criteria, and if the valuer is unaware of which buyers are currently active, and how they determine the price that they can pay for a property, it could adversely affect their ability to provide an accurate valuation.

In *Singer & Friedlander Ltd* v *John D Wood & Co* [1977], Justice Watkins stated in regard to property valuations that:

> The valuation of land by trained, competent and careful professional men is a task which rarely, if ever, admits of precise conclusion. Often beyond certain well-founded facts so many imponderables confront the Valuer that he is obliged to proceed on the basis of assumptions. Therefore, he cannot be faulted for achieving a result which does not admit of some degree of error. Thus, two able and experienced men, each confronted with the same task, might come to different conclusions without any one being justified in saying that either of them has lacked competence and reasonable care, still less integrity, in doing his work.

In this case Justice Watkins introduced the concept of an acceptable margin of error:

> The permissible margin of error is said .... to be generally 10 per cent either side of a figure which can be said to be the right figure, i.e. so I am informed, not a figure which later, with hindsight, proves to be right but which at the time of valuation is the figure which a competent, careful and experienced Valuer arrives at after making all the necessary inquiries and paying proper regard to the then state of the market. In exceptional circumstances the permissible margin, they say, could be extended to about 15 per cent, or a little more, either way. Any valuation falling outside what I shall call the "bracket" brings into question the competence of the Valuer and the sort of care he gave to the task of valuation.

In the DTI report into the collapse of Queens Moat House Hotels in the early 1970s (carried out by Adrian Burn and Patrick Phillips and published in 2004), the specific difficulty of carrying out valuations on hotels was discussed:

> Hotels can differ very significantly from each other in facilities and quality and they can vary greatly in the combination of business activities they attract. Some, for example, offer little other than room hire whilst others

may offer extensive restaurant, leisure and conference facilities. Some are focusing principally on the leisure market, others on the business market.

These, and many other, factors may contribute significantly to the composition and quality of an hotel's profits and their susceptibility to changes in economic conditions.

An appropriate yield, too, is more difficult to assess when valuing a specialised property than it is with non-specialist properties, simply because of the potential volatility of the future profit stream.

Coupled with all this, there are often no useful local comparable transactions and, even when there are, the details of any sale transaction are rarely disclosed. This paucity of information has resulted in some reliance in the hotel market on the very imperfect 'price per room' approach to which we have referred earlier.

Corisand Investments Ltd v *Druce & Co* [1978] is probably the leading case regarding hotel valuations and Mr Justice Gibson was particularly clear in the guidance that was given. He stated that the tests that needed to be applied in this case to determine whether the valuer had failed to do his duty were as follows:

> Are the defendants shown to have produced a valuation in a figure which no competent Valuer could have reached?
>
> Did the defendants ... rely upon any matters upon which no competent Valuer could properly rely, or fail to take into account any matters to which no competent Valuer could in the circumstances fail to have regard?
>
> In particular, in the circumstances of September 1973, could any competent Valuer advising on valuation for mortgage purposes fail to be aware of the implications of the Fire Precautions Act 1971 and to the certain, or at least probable, need for money to be spent on works to put the hotel in compliance with that Act?

Mr Justice Gibson also outlines what steps were required for a valuer to be able to discharge his duty of care in carrying out a valuation. He stated:

> I have no doubt ... that in order to discharge the duty of care of an ordinarily competent Valuer in valuing property such as this hotel the Valuer must have regard to the following matters of principle and of fact:
>
> He must by inspection of the property, and by inquiry, learn enough of the property to be able to start upon the basic method of valuation which he will apply, and thereafter to apply that method effectively by obtaining any further information he needs.
>
> The purpose of the Valuer's work is to determine the price which the property would fetch if offered for sale at the relevant time and in the relevant circumstances.
>
> When he has sufficiently informed himself as to the size, nature and condition of the property he can select the various methods of valuation by which he will guide and check his opinion. For example, he may be able to value by the comparison method — (with or without any other method) — if he has sufficient knowledge of the recent sale prices of other sufficiently comparable properties. It was agreed that the direct comparison method was rarely applicable to hotels, except in some special cases, for example in parts of London where there are a number of hotels which are sufficiently similar for a comparison method to be applied by determining a room price from other sales.

> Hotels are bought and owned to make money by operating them. Accordingly, in estimating what purchasers in the market would pay for a particular hotel, the principal, or at least a well-known and respected, method is to value the hotel as it is as a going concern, including goodwill and contents. The purchaser would calculate what he could expect to earn in the hotel as it stands, or as he could make it operate, and what price it is sensible to pay for the right and opportunity to earn that income. The Valuer tries to make the same calculation.
>
> An experienced Valuer, after inspecting a property, will very frequently if not always readily form an approximate estimate of the probable market price of a hotel. He may test that approximate estimate against the views of people who have immediate knowledge of sale prices in the market. His opportunity to do that will be improved if his own firm has a substantial sale business of hotels.

Mr Justice Gibson went on to say:

> It seems to me that all methods of valuation considered in this case are in truth variations of the comparison method, and all, to a greater or lesser extent, proceed by what has been called, but inaccurately called, the 'instinctive' method. The forming of an 'instinctive' estimate of the value of an hotel must depend upon a Valuer deciding that purchasers are likely to pay for this particular hotel so much more or less than other purchasers paid or offered to pay for other hotels, according to what appears to be the relevant differences in size, location and quality, etc. That is not an instinctive process — it is the making of an estimate of sale value from experience based upon an indirect and general comparison method. It differs from the direct comparison method in one very important particular — the Valuer, if challenged, either does not or cannot produce for scrutiny the comparables upon which he was proceeding in making his estimate. That is no criticism of him — a Valuer acting in his profession will at any date have in mind, no doubt, a large amount of information about the market, much of which he could not recall a year later, and much of which could not be produced for scrutiny.
>
> The accounts analysis method, as it was used by all Valuers who gave evidence in this case, was also and equally a variation of the comparison method without production of comparables for scrutiny. Upon the points in the various accounts analyses upon which there was dispute and discussion, the Valuers were estimating, for example, what the occupancy or the running expenses of the Raglan Hall Hotel would have been by reference to what they believed the occupancy rates and running expenses of other hotels were or were thought to be after making any necessary allowances for relevant differences. Further, when the Valuer settles upon a year's purchase multiplier in his analysis he is deciding what he thinks a purchaser would have paid for the right and opportunity to earn the estimated net annual profit in that hotel by what he knows or believes other purchasers were paying or offering to pay for the right or opportunity to earn similar or different actual or estimated net annual profits in other hotels.

Hotel valuation methodology

There are a number of different basic methodologies that are used across the world to calculate the value of hotels which are outlined below and it is up to the valuer to determine the approach methodology that best reflects the approach taken by the market. However, all methods require as a final step for the valuer to stand back and review the result to ensure it is reasonable.

In this chapter, we will detail the two most important methods of valuation in detail: the profits method of valuation first, then the comparison method. Other methodologies are then outlined afterwards.

Table 6.1 Forecast in present values, Hotel Sarajevo

Year	1		2		3	
No of Rooms	100		100		100	
Rooms Sold	27,375		27,375		27,375	
Rooms Available	36,500		36,500		36,500	
Occupancy (%)	75.0		75.0		75.0	
ADR	80.00		80.00		80.00	
RevPAR	60.00		60.00		60.00	
Growth (RevPAR) (%)			0.0		0.0	
Revenues (£'000s)		(%)		(%)		(%)
Rooms	2,190.0	65.0	2,190.0	65.0	2,190.0	65.0
Food	505.4	15.0	505.4	15.0	505.4	15.0
Beverage	404.3	12.0	404.3	12.0	404.3	12.0
Total Food & Beverage	909.7	27.0	909.7	27.0	909.7	27.0
Leisure Club	202.2	6.0	202.2	6.0	202.2	6.0
Other Income	67.4	2.0	67.4	2.0	67.4	2.0
Total Revenue	**3,369.2**	**100.0**	**3,369.2**	**100.0**	**3,369.2**	**100.0**
Departmental Profit						
Rooms	1,620.6	74.0	1,620.6	74.0	1,620.6	74.0
Total Food & Beverage	363.9	40.0	363.9	40.0	363.9	40.0
Leisure Club	84.9	42.0	84.9	42.0	84.9	42.0
Other Income	23.6	35.0	23.6	35.0	23.6	35.0
Total Departmental Profit	2,093.0	62.1	2,093.0	62.1	2,093.0	62.1
Departmental Costs	1,276.3	37.9	1,276.3	37.9	1,276.3	37.9
Undistributed Operating Expenses						
Administrative & General	286.4	8.5	286.4	8.5	286.4	8.5
Sales & Marketing	134.8	4.0	134.8	4.0	134.8	4.0
Property Operations & Maintenance	134.8	4.0	134.8	4.0	134.8	4.0
Utility Costs	107.8	3.2	107.8	3.2	107.8	3.2
Total Undistributed Expenses	663.7	19.7	663.7	19.7	663.7	19.7
Income Before Fixed Costs	**1,429.2**	**42.4**	**1,429.2**	**42.4**	**1,429.2**	**42.4**
Fixed Costs						
Reserve for Renewals	134.8	4.0	134.8	4.0	134.8	4.0
Property Taxes	158.4	4.7	158.4	4.7	158.4	4.7
Insurance	33.7	1.0	33.7	1.0	33.7	1.0
Management Fees	101.1	3.0	101.1	3.0	101.1	3.0
Total Fixed Costs	427.9	12.7	427.9	12.7	427.9	12.7
EBITDA	1,001.3	29.7	1,001.3	29.7	1,001.3	29.7

Source: Leisure Property Services

The profits method (and the investment method)

There are two main methods for calculating the value of a property using the profits method; the income capitalisation method and a discounted cash flow method.

Method 1: Income capitalisation approach (income cap method)

One of the most common approaches adopted in calculating the value of a hotel is the income capitalisation approach, where the sustainable Earnings before Interest, Taxation, Depreciation and Amortisation (EBITDA) is determined and a multiple is applied to these earnings to determine the value.

These calculations are all in present values, and when the trading has not already stabilised and is not anticipated to stabilise until future years, any growth in the income stream specifically excludes any growth attributable to inflation.

Example 1: Hotel Sarajevo

In the first example, a freehold hotel with the benefit of vacant possession, it has been calculated that the sustainable EBITDA (see Chapter 2) for the 100-bedroom hotel is achievable in year 1 and has been calculated as shown in Table 6.1.

As can be seen, trading is anticipated to remain the same over the three year period, resulting in a RevPAR of £60 in present values, based on a 75% occupancy rate and an £80 ADR.

The revenue mix also remained constant, as did department costs and undistributed operating expenses.

A number of other hotels in the area and of a similar quality have been sold off a capitalisation rate of 9% and it has been determined that this is the appropriate rate to adopt for this hotel (in Chapter 7, yield selection and valuationn multiples are discussed in more detail).

Income capitalisation, Hotel Sarajevo		
Net Base Cash Flow		
EBITDA in 1 values in year 1	1,001,335	
Capitalised at 9.00%	11.11	11,125,949
Gross Value		11,125,949
Say		11,125,000

Source: Leisure Property Services

In simplistic terms, the multiple for a property that has a capitalisation rate of 9.00% (in perpetuity) is 11.11 (1 ÷ 9.00%) or for a fixed term (say for a 45-year lease) it would be 10.88 (1 − (1 + 9%) − 45 ÷ 9%). Parry's tables will happily explain the various permutations of years' purchase (YP) and it is not our intention here to explain the mathematics behind the workings.

In this example, there are no reductions for an income shortfall or capital expenditure, which will be outlined later in the book.

Table 6.2 Forecast in present values, Hotel Vilnius

Year	1		2		3	
No of Rooms	100		100		100	
Rooms Sold	21,900		25,550		27,375	
Rooms Available	36,500		36,500		36,500	
Occupancy (%)	60.0		70.0		75.0	
ADR	65.00		75.00		80.00	
RevPAR	39.00		52.50		60.00	
Growth (RevPAR) (%)			34.6		14.3	
Revenues (€'000s)		(%)		(%)		(%)
Rooms	1,423.5	60.0	1,916.3	63.0	2,190.0	65.0
Food	474.5	20.0	547.5	18.0	572.8	17.0
Beverage	284.7	12.0	334.6	11.0	336.9	10.0
Total Food & Beverage	759.2	32.0	882.1	29.0	909.7	27.0
Leisure Club	142.4	6.0	182.5	6.0	202.2	6.0
Other Income	47.5	2.0	60.8	2.0	67.4	2.0
Total Revenues	**2,372.5**	**100.0**	**3,041.7**	**100.0**	**3,369.2**	**100.0**
Departmental Profit						
Rooms	925.3	65.0	1,370.1	71.5	1,620.6	74.0
Total Food & Beverage	189.8	25.0	308.7	35.0	363.9	40.0
Leisure Club	49.8	35.0	71.2	39.0	84.9	42.0
Other Income	16.6	35.0	21.3	35.0	23.6	35.0
Total Departmental Profit	**1,181.5**	**49.8**	**1,771.3**	**58.2**	**2,093.0**	**62.1**
Departmental Costs	1,191.0	50.2	1,270.4	41.8	1,276.3	37.9
Undistributed Operating Expenses						
Administrative & General	225.4	9.5	273.8	9.0	286.4	8.5
Sales & Marketing	142.4	6.0	152.1	5.0	134.8	4.0
Property Operations & Maintenance	94.9	4.0	121.7	4.0	134.8	4.0
Utility Costs	75.9	3.2	97.3	3.2	107.8	3.2
Total Undistributed Expenses	538.6	22.7	644.8	21.2	663.7	19.7
Income Before Fixed Costs	**642.9**	**27.1**	**1,126.5**	**37.0**	**1,429.2**	**42.4**
Fixed Costs						
Reserve for Renewals	94.9	4.0	121.7	4.0	134.8	4.0
Property Taxes	151.8	6.4	152.1	5.0	158.4	4.7
Insurance	33.2	1.4	33.5	1.1	33.7	1.0
Management Fees	71.2	3.0	91.3	3.0	101.1	3.0
Total Fixed Costs	351.1	14.8	398.5	13.1	427.9	12.7
EBITDA	**291.8**	**12.3**	**728.0**	**23.9**	**1,001.3**	**29.7**

Source: Leisure Property Services

Methodology for Capital Valuations

The value is also reported gross of transaction costs in line with the guidance provided in the *Red Book*.

The mathematical calculation has been rounded down to £11,125,000, which, in this instance, reflects market practice. Once again, whether rounding up or down, the valuer must take into account the prevalent market practice and adopt this when completing the valuation.

However, there will be occasions where the trading of the hotel is not stabilised in the first year; listed below are just a few such reasons:

- The hotel has recently opened and is still building up its trading base.
- The property has recently undergone a comprehensive refurbishment to improve the bedroom product, which will result in an increased performance.
- The proposed purchaser is planning to invest capital expenditure in the property in the future.
- A change in the supply dynamics in the market has occurred, either with competitor hotels closing, or with new hotels opening in the area.
- A change in the demand for hotels in the area, either through new companies opening in the area or changes in the road network or other transport links.
- Changes occurring to the cost structure of the hotel, for example the need to employ additional staff, or a change in kitchen equipment leading to operational savings that need to be reflected.

Example 2: Hotel Vilnius

In this example a very similar hotel to that contained in Example 1 will generate a stabilised EBITDA of 1,000,000, but as a new hotel it will take three years to achieve that level (in present values) as shown in Table 6.2. The property is also a freehold hotel with the benefit of vacant possession.

It is important to state that these projections are in present values, and as such they do not include inflation.

Income capitalisation, Hotel Vilnius

Net Base Cash Flow			
EBITDA in 1 values in year 3		1,001,335	
Capitalised at 9.00%		11.11	11,125,949
Less	Income Shortfall		982,800
Gross Value			10,143,149
Say			10,150,000

Source: Leisure Property Services

The valuation of the property in this example shows the impact of this build-up period until stabilised trading is achieved. As can be seen, we have deducted an income shortfall of €982,800. This is the shortfall in income to the owner over the first two years (until stabilisation) against what would have been earned if the property was earning at its optimum level in the first year.

In this example, the actual calculation is two years of stabilised income less the actual income over the first two years £1,001,300 + £1,001,300 – £728,000 – £291,800 = £982,800.

The same methodology is appropriate for investment properties, as shown in Example 3.

Table 6.3 Forecast in present values, Hotel Reykjavik

Year	1		2		3	
No of Rooms	100		100		100	
Rooms Sold	27,375		27,375		27,375	
Rooms Available	36,500		36,500		36,500	
Occupancy (%)	75.0		75.0		75.0	
ADR	80.00		80.00		80.00	
RevPAR	60.00		60.00		60.00	
Growth (RevPAR) (%)			0.0		0.0	
Revenues (€'000s)		(%)		(%)		(%)
Rooms	2,190.0	65.0	2,190.0	65.0	2,190.0	65.0
Food	505.4	15.0	505.4	15.0	505.4	15.0
Beverage	404.3	12.0	404.3	12.0	404.3	12.0
Total Food & Beverage	909.7	27.0	909.7	27.0	909.7	27.0
Leisure Club	202.2	6.0	202.2	6.0	202.2	6.0
Other Income	67.4	2.0	67.4	2.0	67.4	2.0
Total Revenue	**3,369.2**	**100.0**	**3,369.2**	**100.0**	**3,369.2**	**100.0**
Departmental Profit						
Rooms	1,620.6	74.0	1,620.6	74.0	1,620.6	74.0
Total Food & Beverage	363.9	40.0	363.9	40.0	363.9	40.0
Leisure Club	84.9	42.0	84.9	42.0	84.9	42.0
Other Income	23.6	35.0	23.6	35.0	23.6	35.0
Total Departmental Profit	**2,093.0**	**62.1**	**2,093.0**	**62.1**	**2,093.0**	**62.1**
Departmental Costs	**1,276.3**	**37.9**	**1,276.3**	**37.9**	**1,276.3**	**37.9**
Undistributed Operating Expenses						
Administrative & General	286.4	8.5	286.4	8.5	286.4	8.5
Sales & Marketing	134.8	4.0	134.8	4.0	134.8	4.0
Property Operations & Maintenance	134.8	4.0	134.8	4.0	134.8	4.0
Utility Costs	107.8	3.2	107.8	3.2	107.8	3.2
Total Undistributed Expenses	663.7	19.7	663.7	19.7	663.7	19.7
Income Before Fixed Costs	**1,429.2**	**42.4**	**1,429.2**	**42.4**	**1,429.2**	**42.4**
Fixed Costs						
Reserve for Renewals	134.8	4.0	134.8	4.0	134.8	4.0
Property Taxes	158.4	4.7	158.4	4.7	158.4	4.7
Insurance	33.7	1.0	33.7	1.0	33.7	1.0
Management Fees	101.1	3.0	101.1	3.0	101.1	3.0
Total Fixed Costs	**427.9**	**12.7**	**427.9**	**12.7**	**427.9**	**12.7**
EBITDA	**1,001.3**	**29.7**	**1,001.3**	**29.7**	**1,001.3**	**29.7**
Rent	700.0		700.0		700.0	
Rent as a % of Turnover	20.8		20.8		20.8	
Rent as a % of EBITDAR	69.9		69.9		69.9	
EBITDA	**301.3**	**8.9**	**301.3**	**8.9**	**301.3**	**8.9**

Source: Leisure Property Services

Example 3: Hotel Reykjavik

This example is a freehold hotel, let on a 20-year lease, paying €700,000 per annum to an experienced operator, a level that is thought to reflect the full market rent value of the property (and would be re-lettable at the end of the lease at the same rent).

Although the lease term in this example is 20 years, the €700,000 a year rent can be considered the net income in perpetuity, in current values (ie, the level that the hotel would be considered to be re-let for in 20 years' time).

In this example, we are still concerned with the profitability of the hotel, although for the purposes of the mathematical calculation it seems unconnected. However, as discussed later in Chapter 7, yield selection and valuation multipliers, the choice of the capitalisation rate, will depend on many factors, including the affordability of the rent.

The rent equates to just under 21% of turnover and almost 70% of EBITDAR and the valuer's experience of the local market suggests that the property is fully rented.

We have calculated the EBITDA after the rent has been deducted to be just under 9%, which, when the management fee has been added back, means that the operator will earn approximately 12% of turnover as their reward for operating the hotel.

The capitalisation rate that we have applied is based upon the lease terms, the value of the tenant's covenant in the current market and the affordability of the rent, as we outline in Chapter 7 in more detail.

Income capitalisation, Hotel Reykjavik

Net Base Cash Flow		
EBITDA in 1 values in year 1	700,000	
Capitalised at 6.00%	16.67	11,666,667
Gross Value		11,666,667
Less Transaction Costs	@ 5.75%	634,358
		11,032,309
Say		€11,025,000

Source: Leisure Property Services

It should be noted that, in line with industry practice and RICS guidance, we have deducted transaction costs to arrive at the net value of the investment.

In this instance, the transactions costs have been made up of 4% stamp duty, 1% agents fees (including VAT) and 0.5% legal fees (including VAT). To calculate what these costs are it is important to know that the gross value includes these costs already, and so to work out what the costs actually are the formula is: the gross value ÷ 1 + 5.75%.

Transaction costs vary from country to country and will be determined by the various factors that make up these costs; for example, if stamp duty on property transactions was running at 10%, with agent's fees at 3% (inclusive of VAT) and legal fees at 1% (inclusive of VAT) then the transactions costs would be 14%.

Table 6.4 Forecast in present values, Hotel Valletta

Year	1		2		3	
No of Rooms	100		100		100	
Rooms Sold	27,375		27,375		27,375	
Rooms Available	36,500		36,500		36,500	
Occupancy (%)	75.0		75.0		75.0	
ADR	80.00		80.00		80.00	
RevPAR	60.00		60.00		60.00	
Growth (RevPAR)			0.0%		0.0%	
Revenues (€'000s)		(%)		(%)		(%)
Rooms	2,190.0	65.0	2,190.0	65.0	2,190.0	65.0
Food	505.4	15.0	505.4	15.0	505.4	15.0
Beverage	404.3	12.0	404.3	12.0	404.3	12.0
Total Food & Beverage	909.7	27.0	909.7	27.0	909.7	27.0
Leisure Club	202.2	6.0	202.2	6.0	202.2	6.0
Other Income	67.4	2.0	67.4	2.0	67.4	2.0
Total Revenue	**3,369.2**	**100.0**	**3,369.2**	**100.0**	**3,369.2**	**100.0**
Departmental Profit						
Rooms	1,620.6	74.0	1,620.6	74.0	1,620.6	74.0
Total Food & Beverage	363.9	40.0	363.9	40.0	363.9	40.0
Leisure Club	84.9	42.0	84.9	42.0	84.9	42.0
Other Income	23.6	35.0	23.6	35.0	23.6	35.0
Total Departmental Profit	2,093.0	62.1	2,093.0	62.1	2,093.0	62.1
Departmental Costs	1,276.3	37.9	1,276.3	37.9	1,276.3	37.9
Undistributed Operating Expenses						
Administrative & General	286.4	8.5	286.4	8.5	286.4	8.5
Sales & Marketing	134.8	4.0	134.8	4.0	134.8	4.0
Property Operations & Maintenance	134.8	4.0	134.8	4.0	134.8	4.0
Utility Costs	107.8	3.2	107.8	3.2	107.8	3.2
Total Undistributed Expenses	663.7	19.7	663.7	19.7	663.7	19.7
Income Before Fixed Costs	1,429.2	42.4	1,429.2	42.4	1,429.2	42.4
Fixed Costs						
Reserve for Renewals	134.8	4.0	134.8	4.0	134.8	4.0
Property Taxes	158.4	4.7	158.4	4.7	158.4	4.7
Insurance	33.7	1.0	33.7	1.0	33.7	1.0
Management Fees	101.1	3.0	101.1	3.0	101.1	3.0
Total Fixed Costs	**427.9**	**12.7**	**427.9**	**12.7**	**427.9**	**12.7**
EBITDA	**1,001.3**	**29.7**	**1,001.3**	**29.7**	**1,001.3**	**29.7**
Rent	500.0		700.0		700.0	
Rent as a % of Turnover	14.8		20.8		20.8	
Rent as a % of EBITDAR	49.9		69.9		69.9	
EBITDA	**501.3**	**14.9%**	**301.3**	**8.9**	**301.3**	**8.9**

Source: Leisure Property Services

Example 4: Hotel Valletta (freehold investment property with a rent review due shortly)

The fourth example is a hotel investment similar to the Hotel Reykjavik in example 3, although the rent is reviewed to a market rent every fifth year, and the next rent review is due in one year's time. For the purposes of this example, the current rent passing is €500,000 per annum although the market rent at the current stage is anticipated to be €700,000 per annum (in present values).

As can be seen in the first year of trading (before the rent review), the tenant makes significantly more profit from operating the hotel than after the rent review has been settled.

Income capitalisation, Hotel Valletta

Net Base Cash Flow		
EBITDA in 1 values in year 1	700,000	
Capitalised at 6.00%	16.67	11,666,667
Less		
Income Shortfall		200,000
Gross Value		11,466,667
Less Transaction Costs	@ 5.75%	623,483
		10,843,184
Say		10,850,000

Source: Leisure Property Services

When is the income capitalisation method used?

- It is used for the majority of valuations in two to three-star hotels with vacant possession in the UK. The RICS Trading Related Valuation Group prepared a Valuation Paper (no 6), which became effective in March 2004 called *The Capital and Rental Valuation of Hotels in the UK*. In this paper, they state: "While it is dangerous to over-generalise, the primary method for valuation for most UK commercial hotels at 3 star and below would be on this basis (income capitalisation), as would that for a number of 4 star units as well."
- It is used in conjunction with the DCF method as a check for most four to five-star properties with vacant possession in the UK and Europe.
- It is used for many simple investments (index-linked fixed rents, rather than turnover-related or deals involving future capital expenditure by the landlord), deals in the budget and mid-market sections of the UK market.
- It is used in conjunction with the DCF method for most investment deals in the UK and across Europe.

Table 6.5 Forecast in present values, Hotel Lusaka

Year	1		2		3	
No of Rooms	100		100		100	
Rooms Sold	25,550		26,280		27,010	
Rooms Available	36,500		36,500		36,500	
Occupancy (%)	70.0		72.0		74.0	
ADR	52.50		54.00		56.00	
RevPAR	36.75		38.88		41.44	
Growth (RevPAR) (%)			5.8		6.6	
Revenues (€'000s)		(%)		(%)		(%)
Rooms	1,341.4	60.0	1,419.1	62.0	1,512.6	65.0
Food	447.1	20.0	434.9	19.0	418.9	18.0
Beverage	268.3	12.0	251.8	11.0	232.7	10.0
Total Food & Beverage	715.4	32.0	686.7	30.0	651.6	28.0
Leisure Club	134.1	6.0	137.3	6.0	128.0	5.5
Other Income	44.7	2.0	45.8	2.0	34.9	1.5
Total Revenue	**2,235.6**	**100.0**	**2,288.9**	**100.0**	**2,327.0**	**100.0**
Departmental Profit						
Rooms	912.1	68.0	1,021.8	72.0	1,119.3	74.0
Total Food & Beverage	271.9	38.0	267.8	39.0	260.6	40.0
Room Hire	0.0	100.0	0.0	100.0	0.0	100.0
Leisure Club	53.7	40.0	57.7	42.0	53.8	42.0
Other Income	15.6	35.0	16.0	35.0	12.2	35.0
Total Departmental Profit	1,253.3	56.1	1,363.3	59.6	1,445.9	62.1
Departmental Costs	982.3	43.9	925.6	40.4	881.1	37.9
Undistributed Operating Expenses						
Administrative & General	223.6	10.0	206.0	9.0	197.8	8.5
Sales & Marketing	134.1	6.0	114.4	5.0	93.1	4.0
Property Operations & Maintenance	89.4	4.0	91.6	4.0	93.1	4.0
Utility Costs	71.5	3.2	73.2	3.2	74.5	3.2
Total Undistributed Expenses	518.7	23.2	485.2	21.2	458.4	19.7
Income Before Fixed Costs	**734.6**	**32.9**	**878.0**	**38.4**	**987.5**	**42.4**
Fixed Costs						
Reserve for Renewals	89.4	4.0	91.6	4.0	93.1	4.0
Property Taxes	105.1	4.7	107.6	4.7	109.4	4.7
Insurance	22.4	1.0	22.9	1.0	23.3	1.0
Management Fees	67.1	3.0	68.7	3.0	69.8	3.0
Total Fixed Costs	283.9	12.7	290.7	12.7	295.5	12.7
EBITDA	**450.7**	**20.2**	**587.3**	**25.7**	**691.9**	**29.7**

Source: Leisure Property Services

Method 2: Discounted cash flow approach

A discounted cash flow (DCF) is a projection of future earnings over a period of time to reflect what the expected income will be over the period of the cash flow. It is normal practice in hotel valuations to project earnings over either a 5 or 10-year period. In practice, there are very few occasions when the trading potential of a hotel is expected to grow above inflation after year 5, so there is little difference in the resulting valuation adopting either period.

The DCF approach is similar to the income capitalisation approach, as both take the stabilised EBITDA/income of the property and multiply it by an appropriate rate to calculate the value. The main differences are that inflation/growth is explicitly included within the cash flow (rather than bringing the earnings back to present values) and that the discount rate applied to the stabilised earnings reflects this explicit assumption of growth.

Example 5: Hotel Lusaka (freehold vacant possession hotel)

In this example, a purchaser is looking to buy a property that has been mismanaged and is intending to reposition it in the market, with trading anticipated to stabilise in four years' time.

The net cash flow over the 5 or 10-year period is valued at the appropriate discount rate, and is discounted back depending upon when the income is received. Determining the appropriate discount rate is just as problematic as trying to determine the capitalisation rate to apply in the income capitalisation method. Unless the valuer has evidence at hand of discount rates that have been applied to similar properties the safest course of action is usually to look at the more capitalisation rate evidence that has been used (which is usually more plentiful) and then to adjust it through the use of the explicit inflation rate that has been factored into the cash flow.

For example, if the appropriate capitalisation rate for the property is 9.0% and the cash flow has assumed annual inflation of 2.5%, then the discount rate would be 11.5%.

Cash flow estimates (from profit and loss estimates), Hotel Lusaka

Years	1	2	3	4	5	6	7	8	9	10		
EBITDA (in €'000s)	451	602	727	775	809	829	850	871	893	915	End Value	
Less Capital Expenditure	0	0	0	0	0	0	0	0	0	0		
Net Cash Flow	451	602	727	775	809	829	850	871	893	915	10,166	9.00%

Source: Leisure Property Services

The residual value of the hotel is then calculated assuming that level of income in perpetuity but discounted by 10 years so that the delay in receiving that income is taken into account.

Discounted cash flow, Hotel Lusaka

NPV Net Base Cash Flow	
Discounted at 11.50%	7,642,259
Gross Value	7,642,259
Say	€7,650,000

Source: Leisure Property Services

Table 6.6 Forecast in present values, Hotel St Johns

Year	1		2		3		4		5	
No of Rooms	100		100		100		100		100	
Rooms Sold	25,550		26,280		27,010		27,375		27,375	
Rooms Available	36,500		36,500		36,500		36,500		36,500	
Occupancy (%)	70.0		72.0		74.0		75.0%		75.0%	
ADR	52.50		54.00		56.00		57.50		58.50	
RevPAR	36.75		38.88		41.44		43.13		43.88	
Growth (RevPAR) (%)			5.8		6.6		4.1%		1.7%	
Revenues ('000s)		(%)		(%)		(%)		(%)		(%)
Rooms	1,341.4	60.0	1,419.1	62.0	1,512.6	65.0	1,574.1	65.0	1,601.4	65.0
Food	447.1	20.0	434.9	19.0	418.9	18.0	435.9	18.0	443.5	18.0
Beverage	268.3	12.0	251.8	11.0	232.7	10.0	242.2	10.0	246.4	10.0
Total F&B	715.4	32.0	686.7	30.0	651.6	28.0	678.1	28.0	689.9	28.0
Leisure Club	134.1	6.0	137.3	6.0	128.0	5.5	133.2	5.5	135.5	5.5
Other Income	44.7	2.0	45.8	2.0	34.9	1.5	36.3	1.5	37.0	1.5
Total Revenue	**2,235.6**	**100.0**	**2,288.9**	**100.0**	**2,327.0**	**100.0**	**2,421.6**	**100.0**	**2,463.8**	**100.0**
Departmental Profit										
Rooms	912.1	68.0	1,021.8	72.0	1,119.3	74.0	1,164.8	74.0	1,185.1	74.0
Total F&B	271.9	38.0	267.8	39.0	260.6	40.0	271.2	40.0	275.9	40.0
Leisure Club	53.7	40.0	57.7	42.0	53.8	42.0	55.9	42.0	56.9	42.0
Other Income	15.6	35.0	16.0	35.0	12.2	35.0	12.7	35.0	12.9	35.0
Total Departmental Profit	1,253.3	56.1	1,363.3	59.6	1,445.9	62.1	1,504.7	62.1	1,530.9	62.1
Departmental Costs	982.3	43.9	925.6	40.4	881.1	37.9	917.0	37.9	932.9	37.9
Undistributed Operating Expenses										
Administrative & General	223.6	10.0	206.0	9.0	197.8	8.5	205.8	8.5	209.4	8.5
Sales & Marketing	134.1	6.0	114.4	5.0	93.1	4.0	96.9	4.0	98.6	4.0
Operations	89.4	4.0	91.6	4.0	93.1	4.0	96.9	4.0	98.6	4.0
Utility Costs	71.5	3.2	73.2	3.2	74.5	3.2	77.5	3.2	78.8	3.2
Total Undistributed Expenses	518.7	23.2	485.2	21.2	458.4	19.7	477.1	19.7	485.4	19.7
Income Before Fixed Costs	**734.6**	**32.9**	**878.0**	**38.4**	**987.5**	**42.4**	**1,027.6**	**42.4**	**1,045.5**	**42.4**
Fixed Costs										
Reserve for Renewals	89.4	4.0	91.6	4.0	93.1	4.0	96.9	4.0	98.6	4.0
Property Taxes	105.1	4.7	107.6	4.7	109.4	4.7	113.8	4.7	115.8	4.7
Insurance	22.4	1.0	22.9	1.0	23.3	1.0	24.2	1.0	24.6	1.0
Management Fees	67.1	3.0	68.7	3.0	69.8	3.0	72.6	3.0	73.9	3.0
Total Fixed Costs	**283.9**	**12.7**	**290.7**	**12.7**	**295.5**	**12.7**	**307.5**	**12.7**	**312.9**	**12.7**
Base Rent	250.0	11.2	250.0	10.9	250.0	10.7	250.0	10.3	250.0	10.1
Turnover Rent	242	10.8	253.6	11.1	261.9	11.3	282.8	11.7	292.0	11.9
EBITDA	(41.1)	−1.8	83.8	3.7	180.0	7.7	187.3	7.7	90.6	7.7

Source Leisure Property Services

Methodology for Capital Valuations

Example 6: Hotel St Johns (freehold investment hotel)

In this example, the valuation is of a hotel with an identical trading profile to the Hotel Lusaka but this hotel is let out on a turnover lease, with a fixed base rent of $250,000, and an additional rent based upon the difference between the fixed rent and 22% of turnover. This means that if 22% of turnover equates to $240,000, then the landlord will receive just the base rent, whereas if the turnover rent equates to $280,200, the landlord will receive $28,200 that year.

Cash flow estimates (from profit and loss estimates), Hotel St Johns ($000s)

Years	1	2	3	4	5	6	7	8	9	10	End Value	
Base Rent	250	250	250	250	250	250	250	250	250	250	3,846	6.50%
Additional Turnover Rent	242	266	288	324	348	363	379	394	410	427	5,023	8.50%
Less Capital Expenditure	0	0	0	0	0	0	0	0	0	0		
Net Cash Flow	492	516	538	574	598	613	629	644	660	677		

Source: Leisure Property Services

As can be seen from the above table, the two income streams have been separated and are valued in the table below at different discount rates. This is to differentiate the level of certainty to the investor that the income will be received.

Discounted cash flow, Hotel St Johns

NPV Net Base Cash Flow	
Discounted at 9.00%	3,229,071
Additional Income	
Discounted at 11.00%	2,114,736
Gross Value	5,343,807
Less Costs @ 5.75%	290,562
Net Value	5,053,245
Say	$5,050,000

Source: Leisure Property Services

In this instance, the difference in risk attached to the different income streams has resulted in a 2% differential being applied, hence the 9% discount rate on the base rent and 11% on the additional turnover rent.

Table 6.7 Forecast in present values, Hotel Lilongwe

Year	1		2		3		4		5	
No of Rooms	100		100		100		100		100	
Rooms Sold	25,550		26,280		27,010		27,375		27,375	
Rooms Available	36,500		36,500		36,500		36,500		36,500	
Occupancy (%)	70.0		72.0		74.0		75.0		75.0	
ADR	52.50		55.35		58.84		61.92		64.57	
RevPAR	36.75		39.85		43.54		46.44		48.43	
Growth (RevPAR) (%)			8.4		9.2		6.7%		4.3	
Revenues ('000s)		(%)		(%)		(%)		(%)		(%)
Rooms	1,341.4	60.0	1,454.6	62.0	1,589.1	65.0	1,695.1	65.0	1,767.7	65.0
Food	447.1	20.0	445.8	19.0	440.1	18.0	469.4	18.0	489.5	18.0
Beverage	268.3	12.0	258.1	11.0	244.5	10.0	260.8	10.0	272.0	10.0
Total F&B	715.4	32.0	703.8	30.0	684.5	28.0	730.2	28.0	761.5	28.0
Leisure Club	134.1	6.0	140.8	6.0	134.5	5.5	143.4	5.5	149.6	5.5
Other Income	44.7	2.0	46.9	2.0	36.7	1.5	39.1	1.5	40.8	1.5
Total Revenue	2,235.6	100.0	2,346.1	100.0	2,444.8	100.0	2,607.8	100.0	2,719.5	100.0
Departmental Profit										
Rooms	912.1	68.0	1,047.3	72.0	1,176.0	74.0	1,254.4	74.0	1,308.1	74.0
Total F&B	271.9	38.0	274.5	39.0	273.8	40.0	292.1	40.0	304.6	40.0
Room Hire	0.0	100.0	0.0	100.0	0.0	100.0	0.0	100.0	0.0	100.0
Leisure Club	53.7	40.0	59.1	42.0	56.5	42.0	60.2	42.0	62.8	42.0
Other Income	15.6	35.0	16.4	35.0	12.8	35.0	13.7	35.0	14.3	35.0
Total Departmental Profit	1,253.3	56.1	1,397.4	59.6	1,519.1	62.1	1,620.4	62.1	1,689.8	62.1
Departmental Costs	982.3	43.9	948.8	40.4	925.7	37.9	987.5	37.9	1,029.7	37.9
Undistributed Operating Expenses										
Administrative & General	223.6	10.0	211.2	9.0	207.8	8.5	221.7	8.5	231.2	8.5
Sales & Marketing	134.1	6.0	117.3	5.0	97.8	4.0	104.3	4.0	108.8	4.0
POM	89.4	4.0	93.8	4.0	97.8	4.0	104.3	4.0	108.8	4.0
Utility Costs	71.5	3.2	75.1	3.2	78.2	3.2	83.5	3.2	87.0	3.2
Total Undistributed Expenses	518.7	23.2	497.4	21.2	481.6	19.7	513.7	19.7	535.7	19.7
Income Before Fixed Costs	734.6	32.9	900.0	38.4	1,037.5	42.4	1,106.6	42.4	1,154.0	42.4
Fixed Costs										
Reserve for Renewals	89.4	4.0	93.8	4.0	97.8	4.0	104.3	4.0	108.8	4.0
Property Taxes	105.1	4.7	110.3	4.7	114.9	4.7	122.6	4.7	127.8	4.7
Insurance	22.4	1.0	23.5	1.0	24.4	1.0	26.1	1.0	27.2	1.0
Management Fees	67.1	3.0	70.4	3.0	73.3	3.0	78.2	3.0	81.6	3.0
Total Fixed Costs	283.9	12.7	298.0	12.7	310.5	12.7	331.2	12.7	345.4	12.7
Total Rent Received	626.0	28.0	656.9	28.0	684.5	28.0	730.2	28.0	761.5	28.0
Assessment of Market Rent	400.0	17.9	400.0	17.0	400.0	16.4	400.0	15.3	400.0	14.7
Additional Rent	226.0	10.1	256.9	11.0	284.5	11.6	330.2	12.7	361.5	13.3
EBITDA	(175.3)	−7.8	(54.9)	−2.3	42.4	1.7	45.2	1.7	47.2	1.7

Source: Leisure Property Services

Methodology for Capital Valuations

Example 7: Hotel Lilongwe

In this example, the Hotel Lilongwe is a freehold investment property leased to the tenant on a turnover rent. It has a similar trading profile as the Hotel Lusaka. However in this case the property is over rented.
The lease determines that the tenant will pay rent of 28% of turnover, whereas in this location that is much higher than the market would normally pay for a hotel of this type taking into account the terms of the lease.

In this example, the market rent is assessed to be $400,000 per annum (just under 18% of turnover) and, as such, the actual rental payments have been split into the market rent and additional rent payable under the terms of the lease.

Cash flow estimates (from profit and loss estimates), Hotel Lilongwe ($000s)

Years	1	2	3	4	5	6	7	8	9	10	End Value	
Assessment of												
Market Rent	400	400	400	400	400	400	400	400	400	400	6,154	6.50%
Additional Rent	226	257	285	330	361	381	400	420	441	462	4,396	10.50%
Less Capital Expenditure	0	0	0	0	0	0	0	0	0			
Net Cash Flow	626	657	685	730	761	781	800	820	841	862		

Source: Leisure Property Services

The rent received is taken into account but differing discount rates are applied to the value of that income stream. The logic behind this approach is that should anything happen to the tenant and the property needs to be re-let the market rent is likely to be agreed by a new tenant, whereas there is much less certainty going forward that the "additional rent" would be received.

Discounted cash flow, Hotel Lilongwe

NPV Net Base Cash Flow	
Discounted at 9.00%	5,166,514
Additional Income	
Discounted at 13.00%	2,166,401
Gross Value	7,332,915
Less Costs @ 5.75%	398,716
Net Value	6,934,198
Say	$6,950,000

Source: Leisure Property Services

When are discounted cash flows used?

- Discounted cash flows are widely used by most purchasers and valuers of mid-market and luxury hotels.
- It is commonly used in conjunction with the income capitalisation method for most "branded" hotels.
- It is widely used for all but the simplest investment purchases.

The comparison method — comparable evidence and prices per bedroom

As all valuers are taught as a basic first principle the best evidence of value is that set by the market — unfortunately, in the case of hotels, it is rare that two units are similar enough in their trading patterns and potential to be used as direct comparable evidence.

With that caveat in place, it is still important to state that comparable evidence is extremely important as an indication of the capital value and the yields adopted in the market, and to set a basic tone for the valuation that is being undertaken.

The Valuation Information Paper number 6 entitled *The Capital and Rental Valuation of Hotels in the UK* and prepared by the Trading Related Valuation Group and published in March 2004 by the RICS discusses comparable evidence and its use, as follows:

> The following are to be considered ... comparing one hotel with another:
>
> The comparables — identify these clearly and analyse the subject property.
>
> The comparables may differ in terms of location, facilities, trading records, business mix, operating costs, size of property, trading opportunities, timing of transaction, presence of special purchasers, and so on;
>
> - the existing and potential competition and its impact on trade;
> - the quality of the operation. This may or may not be reflected in its 'star rating', but will be reflected in the size of rooms, the quality of fixtures, fittings, furniture, furnishings and equipment, and the resultant trade;
> - the existence of any franchise or management agreement;
> - conditions attached to any planning permission;
> - the existence of conservation areas and listed building constraints, and the impact of these;
> - the impact of any actual or potential contamination of the property;
> - relevant lease provisions;
> - means of escape in case of fire; and
> - outstanding repairs, maintenance and renewals.

However, in certain circumstances it is commonplace for purchasers to base their acquisition strategy on a price per bedroom, for example in lifestyle properties. A "lifestyle" property is a type of hotel that is usually owned and operated by people who have based their decision to purchase a property on the desire to run a certain type of hotel.

Example 8: Hotel Paris

In example 8, the valuer is being asked to value the Hotel Paris, an 18-bedroom hotel in the Scottish highlands, with a two-bedroom owner's cottage to the rear of the hotel. In the past 12 months, three broadly similar hotels have sold, as detailed below.

- Hotel Sal was located in a comparable village with a similar tourist profile. It has 20 bedrooms and a two-bedroom owner's cottage. It sold nine months ago for £1,600,000, which equates to £80,000 per bedroom.
- Hotel Boa Vista was of a slightly inferior design and in a slightly less desirable location. It had 12 bedrooms and a two-bedroom owner's flat contained within the actual building and sold for £720,000, which equates to £60,000 per bedroom, two months ago.
- Hotel Sao Vincente is the most recent, having sold one month ago. It had 15 bedrooms, a two-bedroom owner's cottage and was in a slightly more desirable location and in better condition. It sold for £1,500,000, which equates to £100,000 per bedroom.

So to summarise:

Table 6.8 Comparing three recently sold hotels

Property	Date sold	No. of bedrooms	Price/bedroom (£)
Hotel Sal	9 months ago	20	80,000
Hotel Boa Vista	2 months ago	12	60,000
Hotel Sao Vincente	1 month ago	15	100,000
Hotel Paris		20	?

Source: Leisure Property Services

The above comparable evidence suggests that the value of Hotel Paris will be somewhere between £60,000 and £100,000 per bedroom (£1,080,000–1,800,000) depending on how the evidence is analysed.

It is essential that the valuer has sufficient experience to be able to accurately analyse the relevant comparable evidence. In this particular example the valuer was able to make the following adjustments:

Hotel Sal

- +5% for the time differential, as the market has strengthened.
- −2% for a slightly smaller property.
- +1% for marginally better condition.
- +0% for location.
  +4% total adjustment.

Hotel Boa Vista

- +0% for time differential.
- +15% for inferior location.
- +5% for smaller property.
- +10% for inferior owner's accommodation.
  +30% total adjustment.

Hotel Sao Vincente

- +0% for time differential.
- +3% for smaller property.
- –8% for inferior location.
- –5% for inferior condition.
  –10% total adjustment.

Table 6.9 Summary of three comparable hotels

	Purchase price/bedroom (£)	Adjustment (%)	Adjusted value/bedroom (£)
Hotel Sal	80,000	+4	83,000
Hotel Boa Vista	60,000	+30	78,000
Hotel Sao Vincente	100,000	–10	90,000

Source: Leisure Property Services

On this basis, the value of the hotel is likely to lie within a range of £78,000–90,000 per bedroom (£1,400,000–1,620,000). The valuer will then be expected to further use his judgment to determine where within that range this particular property would sit. In this instance, it has been determined that the value would be £85,000 per bedroom, equating to £1,530,000.

The appropriate level of adjustment will require a great deal of skill and expertise by the valuer. In each case, the appropriate adjustments will probably differ, depending upon the prevalent market conditions.

Problems with the comparison method

There are two basic problems with relying on comparable evidence; different trading patterns for the particular property and the incomplete detail and general unreliability of the evidence provided.

As a rule of thumb, it is sensible to assume that the nature of the business at every hotel is different, and that the peculiarities of the trading could affect the profitability of the business. As has been shown, it is the level of net income that will have a major impact on the value of a property.

To illustrate this let us assume that we know that a mid-market hotel in Malmo with 50 bedrooms in January 2007 sold for €8,250,000, which equates to €165,000 per bedroom. It is our task to value a mid-market hotel with 150 bedrooms in Copenhagen in March 2007 and the Malmo hotel is the closest geographic comparable that occurred within three months of the valuation date. In theory, therefore, the hotel that we are valuing should be worth around €165,000 a bedroom, at an end value of €24,750,000. However, there are a number of factors that are likely to detract from the accuracy of this assessment, including the following:

- The RevPAR potential of the hotel.
- Size of the property.
- Facilities.
- Costs in the area.
- Choice of capitalisation rate.

Example 9: The RevPAR potential of the hotel

In example 9, Hotel Vienna and Hotel Suva are compared with one another, analysing the trading projections of both hotels (Table 6.10).

The nature of business in Malmo is completely different from that in Copenhagen, with the strong residential conference and group leisure business at Copenhagen supplemented by local corporate demand and individual leisure demand generating substantial RevPARs, based on better occupancies and ADRs.

Taking the projected trading into account (based on the historic accounts), the purchase price of €8,250,000 breaks back to a multiple of just under 14.5 (or a capitalisation rate of 6.9%), as shown in the table below.

Net base cash flow, Hotel Vienna, Malmo

EBITDA in 1 values in year 1	570,276	
Capitalised at 6.90%	14.49	8,264,870
Gross Value		8,264,870
Say		8,250,000
Price per Bedroom		€165,000

Source: Leisure Property Services

Net base cash flow, Hotel Suva, Copenhagen

EBITDA in 1 values in year 1	750,924	
Capitalised at 6.90%	14.49	10,882,961
Gross Value		10,882,961
Say		10,875,000
Price per Bedroom		€217,500

Source: Leisure Property Services

Table 6.10 Forecast in present values, comparing Hotel Vienna and Hotel Suva

	Hotel Vienna, Malmo		Hotel Suva, Copenhagen	
Year	1		1	
No of Rooms	50		50	
Rooms Sold	14,235		14,235	
Rooms Available	18,250		18,250	
Occupancy (%)	78.0		85.0	
ADR	120.00		145.00	
RevPAR	93.60		123.25	
Revenues ('000s)		(%)		(%)
Rooms	1,708.2	65.0	2,249.3	65.0
Food	394.2	15.0	519.1	15.0
Beverage	315.4	12.0	415.3	12.0
Total Food & Beverage	709.6	27.0	934.3	27.0
Leisure Club	157.7	6.0	207.6	6.0
Other Income	52.6	2.0	69.2	2.0
Total Revenue	**2,628.0**	**100.0**	**3,460.5**	**100.0**
Departmental Profit				
Rooms	1,110.3	65.0	1,462.1	65.0
Total Food & Beverage	248.3	35.0	327.0	35.0
Leisure Club	55.2	35.0	72.7	35.0
Other Income	18.4	35.0	24.2	35.0
Total Departmental Profit	**1,432.3**	**54.5**	**1,886.0**	**54.5**
Departmental Costs	**1,195.7**	**45.5**	**1,574.5**	**45.5**
Undistributed Operating Expenses				
Administrative & General	236.5	9.0	311.4	9.0
Sales & Marketing	131.4	5.0	173.0	5.0
Property Operations & Maintenance	105.1	4.0	138.4	4.0
Utility Costs	92.0	3.5	121.1	3.5
Total Undistributed Expenses	**565.0**	**21.5**	**744.0**	**21.5**
Income Before Fixed Costs	**867.2**	**33.0**	**1,142.0**	**33.0**
Fixed Costs				
Reserve for Renewals	105.1	4.0	138.4	4.0
Property Taxes	84.1	3.2	110.7	3.2
Insurance	28.9	1.1	38.1	1.1
Management Fees	78.8	3.0	103.8	3.0
Total Fixed Costs	**297.0**	**11.3**	**391.0**	**11.3**
EBITDA	**570.3**	**21.7**	**750.9**	**21.7**

Source: Leisure Property Services

Methodology for Capital Valuations

However, in the Hotel Suva in Copenhagen the trading profile has only been adjusted to incorporate the increased volume of business at a higher rate than the comparable transaction. Rather unrealistically all the costs associated with the operation of the two hotels have been kept at the same levels (as a percentage of operating income) to show the affect that only these changes will have upon the value of the property.

Taking into account the same yield, it is clear that the value per bedroom is substantially different from the assumed comparable hotel.

However, if we were to assume that the price per bedroom were correct then the value would be €24,750,000. This would represent a capitalisation rate of around 9.1%, which would assume that properties in the better-performing location were worth less than those where the comparable was located.

Example 10: Size of the property

In example 10, the Hotel Kampala, Malmo, which has 150 bedrooms is compared with the Hotel Vienna, Malmo from example 9, which has 50 bedrooms.

The additional bedrooms in Hotel Kampala could lead to higher profitability, as the fixed operational costs are allocated over more bedrooms, thus reducing their impact.

Alternatively, there could be too many bedrooms for the location meaning that the RevPAR achievable at the hotel is lower than should be the case, and thus detracting from the at the hotel's earnings.

In this example, the Hotel Kampala trades at exactly the same occupancy rate and ADR as the Hotel Vienna, but because of the additional bedrooms it has managed to improve its profitability through economies of scale, as shown in the Table 6.11.

Assuming that the same capitalisation rate is applied to the cash flow of the hotel as for the comparable hotel, then the value of the property is approximately 40% higher per bedroom because of the increased efficiencies generated by the more "cost-effective" size of the hotel.

Net base cash flow, Hotel Kampala, Malmo		
EBITDA in 1 values in year 1	2,392,400	
Capitalised at 6.90%	14.49	34,672,461
Gross Value		34,672,461
Say		34,675,000
Price per Bedroom		€231,167

Source: Leisure Property Services

Valuation of Hotels for Investors

Table 6.11 Forecast in present values, Hotel Vienna and Hotel Kampala comparison

	Hotel Vienna, Malmo		Hotel Kampala, Malmo	
Year	1		1	
No of Rooms	50		150	
Rooms Sold	14,235		14,235	
Rooms Available	18,250		18,250	
Occupancy (%)	78.0		78.0	
ADR	120.00		120.00	
RevPAR	93.60		93.60	
Revenues ('000s)		(%)		(%)
Rooms	1,708.2	65.0	5,124.6	65.0
Food	394.2	15.0	1,182.6	15.0
Beverage	315.4	12.0	946.1	12.0
Total Food & Beverage	709.6	27.0	2,128.7	27.0
Leisure Club	157.7	6.0	473.0	6.0
Other Income	52.6	2.0	157.7	2.0
Total Revenue	**2,628.0**	**100.0**	**7,884.0**	**100.0**
Departmental Profit				
Rooms	1,110.3	65.0	3,792.2	74.0
Total Food & Beverage	248.3	35.0	819.5	38.5
Leisure Club	55.2	35.0	177.4	37.5
Other Income	18.4	35.0	55.2	35.0
Total Departmental Profit	**1,432.3**	**54.5**	**4,844.3**	**61.4**
Departmental Costs	**1,195.7**	**45.5**	**3,039.7**	**38.6**
Undistributed Operating Expenses				
Administrative & General	236.5	9.0	662.3	8.4
Sales & Marketing	131.4	5.0	331.1	4.2
Property Operations & Maintenance	105.1	4.0	299.6	3.8
Utility Costs	92.0	3.5	268.1	3.4
Total Undistributed Expenses	**565.0**	**21.5**	**1,561.0**	**19.8**
Income Before Fixed Costs	**867.2**	**33.0**	**3,283.3**	**41.6**
Fixed Costs				
Reserve for Renewals	105.1	4.0	315.4	4.0
Property Taxes	84.1	3.2	252.3	3.2
Insurance	28.9	1.1	86.7	1.1
Management Fees	78.8	3.0	236.5	3.0
Total Fixed Costs	**297.0**	**11.3**	**890.9**	**11.3**
EBITDA	570.3	21.7	2,392.4	30.3

Source: Leisure Property Services

Methodology for Capital Valuations

Facilities

The type of facilities offered at a property is likely to determine the income it can generate, which, in turn, will impact upon the profitability of the hotel. For example, a hotel that relies upon bedrooms for 85% of its total revenue will most likely have a higher proportional EBITDA than one where 55% of its revenue comes from the lower-margin restaurant and bar income, with only 45% coming from the bedrooms.

Example 11: Hotel Nicosia

In example 11, the Hotel Nicosia in Copenhagen has only a small restaurant and bar, with some meeting rooms and no leisure facilities. There are also four areas that have been let out to third-party operators (including a large restaurant and a health and fitness centre) where the hotel receives a fixed rental income.

Once again, we have unrealistically kept all the unallocated costs the same as for the Hotel Vienna (with the additional cost-intensive facilities), rather than reducing them as would likely be the case, to demonstrate the impact of the changing business mix alone.

Net base cash flow, Hotel Nicosia, Copenhagen		
EBITDA in 1 values in year 1	810,390	
Capitalised at 6.90%	14.49	11,744,785
Gross Value		11,744,785
Say		11,750,000
Price per Bedroom		€235,000

Source: Leisure Property Services

Once again, the impact of the change in business mix at the hotel has been substantial on the value per bedroom of the hotel, some 38% higher than the comparable rate (assuming the same capitalisation rate).

Costs in the area

The level of costs associated with the hotel will obviously affect the bottom line profitability and will, therefore, render a straightforward bedroom comparison inaccurate. This is especially important in areas where employee rights are strong and where labour unions are powerful. In France, for example, the labour costs associated with a new hotel (where all the employees are newly employed) can be 15–25% lower than in an existing hotel where overmanning has occurred "organically".

Valuation of Hotels for Investors

Table 6.12 Forecast in present values, Hotel Vienna and Hotel Nicosia comparison

	Hotel Vienna, Malmo		Hotel Nicosia, Malmo	
Year	1		1	
No of Rooms	50		50	
Rooms Sold	14,235		14,235	
Rooms Available	18,250		18,250	
Occupancy (%)	78.0		78.0	
ADR	120.00		120.00	
RevPAR	93.60		93.60	
Revenues ('000s)		(%)		(%)
Rooms	1,708.2	65.0	1,708.2	85.0
Food	394.2	15.0	40.2	2.0
Beverage	315.4	12.0	40.2	2.0
Total Food & Beverage	709.6	27.0	80.4	4.0
Leisure Club	157.7	6.0	0.0	0.0
Other Income	52.6	2.0	221.1	11.0
Total Revenue	**2,628.0**	**100.0**	**2,009.6**	**100.0**
Departmental Profit				
Rooms	1,110.3	65.0	1,238.4	72.5
Total Food & Beverage	248.3	35.0	32.2	40.0
Leisure Club	55.2	35.0	0.0	0.0
Other Income	18.4	35.0	199.0	90.0
Total Departmental Profit	**1,432.3**	**54.5**	**1,469.6**	**73.1**
Departmental Costs	1,195.7	45.5	540.1	26.9
Undistributed Operating Expenses				
Administrative & General	236.5	9.0	180.9	9.0
Sales & Marketing	131.4	5.0	100.5	5.0
Property Operations & Maintenance	105.1	4.0	80.4	4.0
Utility Costs	92.0	3.5	70.3	3.5
Total Undistributed Expenses	**565.0**	**21.5**	**432.1**	**21.5**
Income Before Fixed Costs	**867.2**	**33.0**	**1,037.5**	**51.6**
Fixed Costs				
Reserve for Renewals	105.1	4.0	80.4	4.0
Property Taxes	84.1	3.2	64.3	3.2
Insurance	28.9	1.1	22.1	1.1
Management Fees	78.8	3.0	60.3	3.0
Total Fixed Costs	**297.0**	**11.3**	**227.1**	**11.3**
EBITDA	**570.3**	**21.7**	**810.4**	**40.3**

Source: Leisure Property Services

Choice of capitalisation rates

In our example, Copenhagen is considered substantially more desirable from both the operator's and owner's perspectives, and as such they would apply a sharper cap rate (a higher earnings multiples) to buy the property.

When is the comparison method used most often?

The RICS working paper for hotel valuations states that prices per bedroom are considered to be at their most useful when valuing "lifestyle" types of hotels or "trophy" hotels.

Trophy hotels are properties that are purchased in part because of the kudos of owning the property, leading to some transactions looking very expensive when analysed using normal capitalisation/discount rates.

It is true that reference will be made to price/bedroom when valuing both lifestyle and trophy hotels and that when there is a dearth of yield evidence in the market such comparables may well influence the capitalisation rate adopted when undertaking the valuation.

Check methods, and determining factors in the purchase of hotels

There are various other methods used to provide a "rule of thumb" check on valuations, including multiple of earnings, multiple of ADRs, internal rate of returns and net initial yields.

Very few of these methods are ever used as anything other than to check the end result of other valuation methods, although we include them here as a valuer or purchaser may come across the methodology and would need to understand how they work.

Multiple of earnings

In the past, and in certain property markets some types of hotel tended to trade on a multiple of earnings; should buyers in any market still be predominantly purchasing properties based on this method then that should be the methodology used, although its use seems to be less prevalent in recent years where buyers have tended to move towards a more sophisticated methodology when assessing the pricing of the property.

Example 12: Hotel Cagliari

In example 12, Hotel Cagliari, the hotel had earned €3,000,000 turnover in the year of its sale with vacant possession, and if the typical market multiple was 3 times earnings, then the hotel would be worth around €9,000,000.

$$(€3,000,000 \times 3 = €9,000,000)$$

Multiple of ADR

A number of people use a simple rule of thumb to check the end calculations of value that is relatively straightforward to apply, although it is not recommended to use it as anything other than as a health check on the formal valuation process, as a number of factors could seriously affect the accuracy of this method.

It is suggested that the ADR of a property multiplied by 1,000 and the number of bedrooms in the hotel will show the value of the property.

Example 13: Hotel Palermo

In example 13, the Hotel Palermo is being sold with vacant possession. The hotel has an ADR of €90.00 and has 100 bedrooms then the property would be worth €9,000,000.

$$(€90 \times 1,000 \times 100)$$

Internal rate of return (IRR)

Some purchasers will not consider purchasing a property unless it has achieved a certain internal rate of return (IRR) and this can influence their purchasing decision. Although not used as a method of calculating the value of a property, it can be used as an overriding factor in the decision-making process. If the IRRs are too low certain investors will not invest, and if the IRRs are higher than the target hurdle rate, investors may consider paying even more if required to secure the property.

The glossary of property terms defines the IRR as: "the rate of interest at which all future cash flows must be discounted in order that the net present value of those cash flows, including the initial investment, should be equal to zero".

Certain investors have specific investment criteria, some of which are to meet specific IRRs. In these instances, the IRR becomes quite important in the process of deciding whether to invest in a certain property and, as such, can influence the value of a property if the majority of potential purchasers require certain IRRs.

Example 14: Hotel Brussels

It is proposed to purchase the hotel for £10.54m and with 5% costs on top, this will result in an initial investment of £11.07m. To calculate the IRR, the net present value of the cash flow is calculated as below. The 10th year's income includes the notional sale value of the property as well as the income from that year.

The most significant problem with the IRR calculation is that the formula assumes that you are reinvesting the annual cash flow at the same rate as calculated by the IRR. As a result, when you have a property that generates significant cash flow, the calculated IRR will overstate the likely financial return of the property.

The modified internal rate of return (MIRR) allows the purchaser to enter a different rate that is applied to the property's annual cash flow. This rate used is generally a bank or savings rate. Using the MIRR will more closely mimic reality, as you are rarely able to reinvest the cash flow at the same rate of return as determined by the IRR formula.

Methodology for Capital Valuations

Table 6.14 Forecast in present values, Hotel Brussels

Year	1		2		3		4		5	
No of Rooms	100		100		100		100		100	
Rooms Sold	27,375		27,375		27,375		27,375		27,375	
Rooms Available	36,500		36,500		36,500		36,500		36,500	
Occupancy (%)	75.0		75.0		75.0		75.0		75.0%	
ADR	80.00		82.00		84.05		86.15		88.31	
RevPAR	60.00		61.50		63.04		64.61		66.23	
		(%)		(%)		(%)		(%)		(%)
Total Revenues	3,369.2	100.0	3,453.5	100.0	3,539.8	100.0	3,628.3	100.0	3,719.0	100.0
Income Before Fixed Costs	1,429.2	42.4	1,465.0	42.4	1,501.6	42.4	1,539.1	42.4	1,577.6	42.4
Total Fixed Costs	427.9	12.7	438.6	12.7	449.6	12.7	460.8	12.7	472.3	12.7
EBITDA	1,001.3	29.7	1,026.4	29.7	1,052.0	29.7	1,078.3	29.7	1,105.3	29.7
Year	6		7		8		9		10	
No of Rooms	100		100		100		100		100	
Rooms Sold	27,375		27,375		27,375		27,375		27,375	
Rooms Available	36,500		36,500		36,500		36,500		36,500	
Occupancy (%)	75.0		75.0		75.0		75.0		75.0%	
ADR	90.51		92.78		95.09		97.47		99.91	
RevPAR	67.88		69.58		71.32		73.10		74.93	
Growth (RevPAR) (%)	2.5		2.5		2.5		2.5		2.5	
		(%)		(%)		(%)		(%)		(%)
Total Revenues	3,812.0	100.0	3,907.3	100.0	4,005.0	100.0	4,105.1	100.0	4,207.7	100.0
Income Before Fixed Costs	1,617.0	42.4	1,657.5	42.4	1,698.9	42.4	1,741.4	42.4	1,784.9	42.4
Total Fixed Costs	484.1	12.7	496.2	12.7	508.6	12.7	521.3	12.7	534.4	12.7
Incentive Management Fee	0.0	0.0	0.0	0.0	0.0	0.0	0.0	0.0	0.0	0.0
EBITDA	1,132.9	29.7	1,161.2	29.7	1,190.3	29.7	1,220.0	29.7	1,250.5	29.7

Source: Leisure Property Services

Internal rate of return, Hotel Brussels

Gross Purchase Price		11,070,000
Net Cash Flow	1	1,001,335
	2	1,026,369
	3	1,052,028
	4	1,078,329
	5	1,105,287
	6	1,132,919
	7	1,161,242
	8	1,190,273
	9	1,220,030
Including End Value	10	15,145,316
IRR		11.41%

Source: Leisure Property Services

Net initial yield (NIY)

Some purchasers will not consider purchasing a property unless it will achieve a certain net initial yield (NIY). Although not used as a method of calculating the value of a property it can certainly be used as an overriding factor in the decision-making process. If the NIY is too low, certain buyers will not invest.

The NIY is the initial net income at the date of purchase expressed as a percentage of the gross purchase price including the costs of purchase. In example 14, for the Hotel Brussels the NIY would be 9.05%

$$\frac{£1,001,335}{£11,070,000} = 9.05\%$$

Valuations in exceptional circumstances

In some circumstances, carrying out a valuation of a hotel can be especially problematic, for example, at the time of a property crash or a property boom, or in periods of uncertainty such as just after the terrorist attacks on 11 September 2001.

In such times, the particular circumstances and how they impact on the hotel market can make it extremely difficult to assess the value of a trading entity such as a hotel and therefore to assess the value. During the early 1970s when the hotel market was in strong decline with almost no open market transactions occurring, it was almost possible for the valuer to report values based on what hotels were not selling for, due to the dearth of transactional evidence.

It is vital that the valuer clarifies to whoever has instructed them any problems that are present in the market that could cause uncertainty over the valuation. In *Merivale Moore plc* v *Strutt & Parker* [1999] the judge stated:

> I find that (the Valuer) did not qualify his advice on yields in any way. He did not warn that there was no established market or that his figure for the yields carried an enhanced risk as compared with figures for freehold, long leasehold and short term leasehold. He did not warn that the yield chosen was based on opinion and not on evidence. I find that the plaintiffs relied upon (the Valuer's) appraisal both in respect of rental income and yield. If they had thought that the figure for rental income in the appraisal did not justify the purchase price, including a 15% profit, they would not have bought the property. If (the Valuer) had qualified his advice on yields in any way, they would not have bought the property.

In the DTI report into the collapse of Queens Moat House, when discussing the valuation of hotels at the deepest part of the 1973 property recession, they mentioned the use of open market value for existing use as an assessment of value for balance sheet purposes.

> This may well have worked satisfactorily in a normal market when there were plenty of willing buyers and willing sellers around to establish the level of the market but what happened when the market was in deep recession as it was in the first half of 1993? Potential buyers had withdrawn from the market to await its recovery or further fall, so what prices were available were such that no owner was going to be a willing seller until a recovery emerged. As a result there was an inadequate amount of information for the Valuer to gauge the level of the market. In these circumstances when there was no 'open market' there could be no reliable open market values.

The report went on to say:

> (valuations) . . . are far more difficult in times of a boom where buyers are falling over themselves to buy at almost any price, or in deep recession when no one not under pressure to sell would ever consider selling.

Summary

In summary, the valuer must adopt the prevalent methodology of the active buyers in the local hotel market; to do that the valuer must be aware of who is actively buying hotels, and how they determine the price that they will pay.

Assessing the value in times of extreme growth or recession is equally difficult. The valuer must understand the cyclical nature of the hotel market and the external factors that influence both trading and acquisition decisions.

It is considered good practice to undertake both profits method and comparable method valuations and to review the results of both before determining the final valuation.

Yield Selection and Valuation Multipliers

Introduction

In this chapter, we outline a number of things that affect yield choice, the appropriate yield or valuation multiplier that is applied to the stabilised earnings to determine the value of the property. This chapter has been subdivided into vacant possession and investment properties for ease of reference.

Determining the correct capitalisation rate and discount rate with which to value a property is fundamental in producing an accurate valuation. It is essential that every reference possible be made to market evidence, but there are many problems with such evidence in the context of the individual (and indeed often secretive) nature of hotel transactions.

The market generally discusses the yield off which a property was purchased. This is calculated by taking the income (whether rent or stabilised EBITDA) and dividing it by the purchase price (and in the case of investments deducting the purchaser's costs).

Example 15: Hotel Moscow

The Hotel Moscow produces $1,200,000 EBITDA (stabilised trading) and was purchased with vacant possession for $12m. The yield paid for the property was 10%.

$$\frac{1,200,000}{12,000,000} = 10\%$$

It is also important to remember that when we are discussing the appropriate yield to adopt we are trying to look into the mind of the potential purchaser. The only way to determine what the "reasonably efficient operator" is likely to pay is to consider who the bidders are likely to be for a particular property and look into the factors that could affect their bid price.

Terminology used in the hotel industry in respect of yields/discount rates and multipliers is similar to the rest of the property market. The appropriate multiplier is calculated by the following 1 ÷ yield. For example, if the hotelier is prepared to pay 8%, then the multiplier they will adopt is 12.5 times

$$\frac{1}{8\%} = 12.5$$

In this section, when discussing the appropriate capitalisation rates that should be adopted in a valuation, we have used "softer" and "sharper" in the place of "higher" or "lower". The use of the latter can be confusing, as a higher capitalisation rate might suggest a more desirable property, whereas it can actually mean the opposite.

Vacant possession properties

The appropriate multiplier is the one that the successful purchaser would adopt when purchasing that specific property. It is likely that the market will consider many factors when deciding what multiple of earnings to pay, including the following:

- *Geographic location.* The location of a property, in terms of its surroundings, is obviously fundamental. The capitalisation rate for a hotel on the Champs Elysee in Paris is likely to be sharper than the capitalisation rate used for a hotel in Freedom Square in Tblisis. It is also likely that the yield will be sharper in a city centre rather than in a secondary roadside location.

- *Suitability of location.* If the property is located in a area to which it is not suited, then the yield is likely to be softer than if it is well located. For example, a conference hotel in a remote location may not attract as sharp a yield as one located in a highly visible location in the centre of a number of motorway connections.

- *Type of property.* Whether the property is a deluxe five-star hotel or a bed & breakfast-style of operation will have an impact on the yield that will be applied. However, it should be noted that it is not a straight forward descending scale with the sharpest yields being applied to the "higher-graded" establishments, as the capitalisation rate will depend upon the market demand for that category of property

- *Quality and condition.* The quality of the property may impact on the yield, as a hotel that is in better condition and requires less defensive capital expenditure will usually command a sharper yield than a run-down equivalent property.

- *Market expectations and perceived desirability in the current market.* The underlying market sentiments towards a property will impact on the yield that property can command. For example, if the "market" feels that mid-market hotels in Dubrovnik are under pressure from newer, boutique hotels and more budget operations, then potential purchasers will not be prepared to pay as much for such a unit, unless it has development potential.

- *Current and proposed competition.* The strength of demand for a particular property will have a fundamental impact on the market yield. If many purchasers are after a particular type of property in a specific location, then a higher price will have to be paid to beat off the challenge of other purchasers.

- *Trading demand.* If the hotel market is particularly buoyant, then the number of new entrants into the market trying to gain a foothold will increase competition for the property, leading to a sharpening of the yield.

- *Historic trading profile.* The historic trading profile of the property will have an impact upon the yield adopted. In many cases, the more stable the trading history (and therefore the less risk associated with this element of the valuation) the sharper the yield profile adopted. However, where there is growth potential in the trading (for example, the hotel has been underperforming and the purchaser believes that they can improve trade), then a sharper yield may be applied to reflect this growth potential.

- *Future growth in value.* If an underlying growth in value of the property is likely (for example, if the country has just been accepted in the European Union and demand for that perceived product is likely to grow substantially), then a sharpening of the yield is likely to reflect this future residual value growth.

- *Business mix and the risk attached to various income streams.* Arguably if a substantial proportion of revenue comes from a less secure income stream (for example, Food & Beverage or short-term sub-lettings rather than from Rooms), then the overall capitalisation rate that it used will be slightly softer to reflect the additional risk to the projected income.

- *Changes in the economic environment.* If an area's economic profile is changing, then this could affect the appropriate capitalisation rate.

- *Changes in the local infrastructure.* Any changes that could affect future trading are likely to be partially reflected in the selection of the capitalisation rate.

- *Alternative uses.* Any alternative uses that could improve the value of the property in the future are likely to be reflected by the selection of the yield.

- *Alternative properties.* If the property is the only option that an operator has in the local area then the yield will be sharper, but if there are many alternative properties of a similar suitability then a lower multiplier is likely to be adjusted.

- *"Treasure hunting" and releasing additional value.* It is likely that if there are opportunities to improve the trading of the property through redevelopment, extensions or repositioning in the market place, then the purchaser will offer a sharper yield to secure the property.

- *Visibility and accessibility of the property.* When visibility is important to the operation of the hotel, a higher-profile outlook will often command sharper yields. Alternatively, if a property relies upon tourists for its custom and there is only one flight a week to the destination, the capitalisation rate may be lower than an equivalent property served by 20 flights a week.

- *Suitability for current use.* If the property is not suited to the use to which it is being purchased for, then a softening of the yield will reflect this.

- *Flexibility.* The more flexibility that the property has can be a commercial advantage for the operator and this could result in a sharper yield being paid.

- *Brandability.* If a property is run independently but is brandable then it could be that a higher proportion of people will be interested in buying the property (potential franchisees for example), which could lead to more competition for the property and therefore a sharper yield.

- *Funding.* The availability of funding and the provision of debt can have an impact on yields. Where high debt levels are freely available yields are likely to be sharper than when debt is limited and more equity is required to fund the purchase.

The above points are just a selection of factors that could have an impact on the yield or multiplier used when assessing the value of a hotel. It should be noted that everything that could affect the profitability of the hotel and/or future growth in the capital value is likely to have an impact on the appropriate yield.

Yield selection for investments

In the hotel world there is no set formulaic standard for leases or management contracts that can be compared with the institutional leases that are commonplace in retail or commercial investments: As such, each investment needs to be assessed on an individual basis.

Most of the criteria that will be considered (location, quality, economic environment etc) for VP properties will also apply when adding the appropriate yield for investment properties but, in addition, there will be other factors to consider, including the following:

- **Who is the tenant (or manager in the case of a management contract)?** The tenant will have a major impact on the yield that will be applied to an investment. If the same property is let to an international hotel operator with good financial standing, then it is likely that more people will be prepared to buy the investment because of the perceived lower risk profile pushing up its desirability and therefore the price (hence sharpening the yield) than if it were let to a local operator with no experience and no money.

    The financial standing of a company is often reviewed through various rating agencies, such as Dunn & Bradstreet or Standard & Poors, and they provide risk assessments that the investment market looks to when assessing how desirable they will be as a tenant.

    Other factors, such as the quality of the operator, their suitability for the operation, the track record of the operator, the stated goals of the operator and the market perception of the operator, will also have an effect on the market yield for such an investment.

- **Is there a guarantor to the rent and are they reliable?** When the financial standing of a tenant is not as strong as an investor would like either the price of the asset is reduced (yield softens) to reflect the additional risk, or a guarantee can be provided. In many countries, this can be a sum of money placed on deposit in a bank with the investor being able to take this money in the event of default. In other cases (and more common in the UK), the provision of a guarantee from an entity with a greater financial standing is provided.

- **Is the rent affordable?** Whether the rent can be paid by the tenant is a key consideration of what yield should be applied to a property. If the rent is too high, there is a much higher likelihood of default and the price paid is likely to reflect this additional risk (and the yield should soften). Conversely if the rent is substantially under the market level (as in the case of a ground rent), the yield will sharpen to reflect the additional certainty of receiving the rent.

As such, if a new hotel is being built it may be let to a specifically created company but it may have to be guaranteed by the parent company to provide the financial backing to increase its value.

Example 16: Hotel Dar es Salaam

In this example, Hotel Dar es Salaam is being let to a tenant at £900,000pa, annually index linked. Such leases to the same tenant normally sell, if the rent is at an acceptable level (compared with market levels), at 6%, which would suggest a gross value of £15,000,000.

However, in this case the EBITDA of the property is £1,100,000 and, as such, the rent equates to 81% of EBITDA. This is considered by the market to be potentially too high, as a slight downturn in the market could lead to a reduction in the EBITDA and leading the tenant into financial troubles. This level of over-rentedness could lead the value either to apply a softer yield on all of the rent (in this case 6.5%, leading to a value of £13,850,000) or to value the market rent at 6% and the over-rented element of a softer yield (6% on the first £650,000 and 15% on the "over-rented" £450,000, resulting in a value of £13,790,000).

Example 17: Hotel Basseterre

In this example, the rent for Hotel Basseterre has been agreed at €350,000 and the EBITDA generated at the hotel is €1,250,000. Other investments of a similar nature with the same tenant would normally command a yield of 7%, suggesting a gross value of €5,000,000 1 ÷ 7% = 14.286, 14.286 × €350,000 = €5,000,100, rounded down to €5,000,000. In this case, the rent is lower than would normally be considered a market rent (at 31% of EBITDA) and so an investor is likely to be prepared to pay more money for the investment as the tenant is unlikely to default and if they do the property is likely to be easily re-let at a higher rent.

As such, the appropriate yield is sharper than the market 7% based on fully rented properties and it has been assessed at 5.5%, showing a gross value of €6,365,000 1 ÷ 5.5% = 18.18, 18.18 × €350,000 = €6,363,636, rounded up to €6,365,000.

Impact of specific lease terms on the adopted yield

Almost every lease term will have an impact on the appropriate yield for an investment, as every term will affect the desirability and the value of the investment. It is not our intention to discuss all of these, but we comment on a few key clauses below:

- There are many different types of leases that calculate the rent in various different ways, for example:

    - Five-yearly reviews to market rent.
    - Fixed rent with fixed uplifts.
    - Fixed rent with annual uplifts based on RPI.

- Turnover-related rents.
- EBITDA-based rents.

The differing levels of certainty of the rental income needs to be reflected in the adopted yields.

- Unusual review periods. Where leases provide rent reviews at different intervals from those being agreed for new lettings in the open market (some markets are annually reviewed while others receive five-yearly reviews), it can sometimes have an effect on the yield applied to the investment.

- User covenants. Where a user clause is qualified to the effect that the use cannot be changed "without consent, such consent not to be unreasonably withheld", there is unlikely to be a large impact on the yield adopted for the valuation.
  Where the prohibition on change of use is absolute, this can effect the value. A restriction to a generic use such as "hotel" will not normally have any valuation consequences. Absolute prohibitions that narrow the range of potential demand can adversely affect rental value (for example, if the clause specifies a use as a "designer hotel"). *Plinth Property Investments Ltd v Mott, Hay and Anderson* (1979) was one of the earliest cases in which the adverse effect of a restricted use was recognised. The court upheld the arbitrator's award for a discount in excess of 30% for a restriction on an office building in Croydon to be used for "consulting engineers".

- Restrictions on alienation. The ability for tenants to be able to assign or sublet their lease is important, and a complete prohibition against alienation can have a severe effect on value. A hotel with a prohibition on subletting may also result in a discount to the value if this restricts the flexibility of the tenant to let out surplus space to complementary uses.

- Alterations. A prohibition on alterations, or the requirement for the landlord's consent (the granting of which can be unilaterally withheld) may impact on the desirability of the investment and hence the yield.

- Security of the rent. How the rent is calculated and therefore how secure it is will have an effect on the value of the investment.

Example 18: Hotel Belfast

In this example, the hotel investor's rent comes from different sources. The investor receives a base rent of £500,000 per year, which is annually reviewed to RPI. There is also a surplus rent received that is based on a turnover top-up. The rent has been set at 22% of total revenue, and if the turnover rent exceeds the base rent, then this additional rent will be received. If the turnover rent is less than the base rent, then the full amount of the base rent will be received but no surplus rent will be received.

The rent is effectively upward and downward, although only one part of the rent (the surplus rent) can reduce. The level of certainty attached to this rent is much less than if it were attached to the base rent and the most sensible way to value these different income streams is off two separate yields to reflect the degree of risk.

Yield Selection and Valuation Multipliers

Table 7.1 Rental income, Hotel Belfast

Years	1	2	3	4	5
Turnover (£000's)	2,850	2,936	3,024	3,114	3,208
Total Rent (t/o %)	22.0	22.0	22.0	22.0	22.0
Total Income (£000's)	627	646	665	685	706
Guaranteed (£000's)	550	564	578	592	607
Surplus (£000's)	77	82	87	93	99

Source: Leisure Property Services

Table 7.2 Discounted cash flow, Hotel Belfast

Guaranteed Income Discounted at 7.50%		10,833,366
Surplus Income Discounted at 10.00%		1,651,235
Gross Value		12,484,601
Less Transaction Costs	5.75%	678,832
Net Value		11,805,769
Say		€11,800,000

Source: Leisure Property Services

Example 19: Hotel Victoria

In this example, the hotel operator pays a rent based entirely upon turnover. Hotel Victoria has been trading for 10 years and has a well-established market segment and a loyal client base. However, the rent will still be subject to the strengths and weaknesses of the market, and indeed changes to supply and demand.

As such, a valuer would have two routes to undertake the valuation. The first method would be to apply one yield across the whole income stream, reflecting the market's belief in the hotel's projections and the strength of that local market.

Table 7.3 Income capitalisation, method 1, Hotel Victoria

Guaranteed		
Income in present values in year 1	627,000	
Capitalised at 7.1%	14.085	8,830,986
Gross Value		8,830,986
Less Transaction Costs	5.75%	480,172
Net Value		8,350,814
Say		€8,350,000

Source: Leisure Property Services

Valuation of Hotels for Investors

However, that is quite difficult to do, as there is unlikely to be any direct evidence.

The second method would be to top slice the rent once again, assuming that certain elements within the turnover rent are likely to be pretty secure, with the last part of the rent being more at risk and therefore at a softer yield.

Table 7.4 Income capitalisation, method 2, Hotel Victoria

Guaranteed		
Secure Rent in present values in year 1	475,000	
Capitalised at 6.50%	15.385	7,307,692
Surplus		
Surplus in present values in year 1	152,000	
Capitalised at 9.85%	10.152	1,543,147
Gross Value		8,850,840
Less Transaction Costs	5.75%	481,251
Net Value		8,369,588
Say		€8,370,000

Source: Leisure Property Services

- Proximity to vacant possession value. A key comfort for most investors is the vacant possession value of a hotel. If the tenant defaults on the rent, what price could be achieved if the property is placed on the market with vacant possession (VP). If the vacant possession value is substantially higher or lower than the investment price, it may have an effect on the appropriate yield to apply to that property.

Example 20: The Hotel Canberra

In this example, Hotel Canberra is worth £10,000,000 with vacant possession. The tenant on similar terms would normally be purchased using a 6.5% capitalisation rate. The rent they are paying is £850,000, which suggests a gross value of £13,000,000, some 30% higher than the VP value.

This is likely to be considered to be too large a difference between the investment value and the VP value. In this instance, the difference could be because the rent is unaffordable, or it could be that not sufficient weight has been placed on the location of the property (because if the rent is affordable then the VP yield is quite soft, suggesting it is not a desirable location in which to have a hotel).

- Useful economic life. If the building is of non-traditional design with a specific useful life and that lifespan is coming close to the end then the yield that is applied to the investment is likely to be relatively softer than if traditional construction methods had been used and there is a reasonable unexpired useful lifespan remaining for the building.

- Condition of the asset. Although it may be a full repairing and insuring lease the condition of the asset will affect the choice of capitalisation rates. The condition of the property will be fundamental to the choice of yield if the investment is a management contract, as it will directly affect the income received from the investment.

- Quality of the management. This is more important for variable leases or management contract investments, although it is important for traditional fixed rent investments as it can affect the residual value of the investment.

- Brand. Certain brands are considered significantly more attractive than others and, as such, command sharper yields.

- Quality of the CRS/GDS. The perceived quality of the central reservations system or global distribution system will have an effect on the perceived desirability of the investment, especially in turnover leases or management contracts.

- Market positioning. Certain hotel investments benefit from (or are hindered by) their status within the relevant hotel market and this can affect the relevant market capitalisation rates that would be adopted.

The key terms for each type of investment

To be able to undertake an accurate valuation of a hotel investment (whether lease or management contract) it is essential that the valuer has a clear understanding of the fundamentals of the market and the investment structure.

In this section, we outline some of the key elements of each type of investment.

Ground leases

A ground lease is a lease where the undeveloped land is leased, usually for a long time. The key terms include:

- The term. The length of the lease will impact upon the value of the lease to both parties; if it is too short the operator will have a shorter period in which to recoup the building costs and so will not be able to pay as high a rent and if it is too long the investor's residual value (when they can regain possession of the land) will be low, suggesting a higher rent (or initial premium) will need to be paid.

- The demise. What is included within the lease will have an effect on the value of the lease. If the land extends to a significantly larger plot than the hotel building actually occupies, it may be that in the future this land will have additional value; as such the tenant may be prepared to pay a higher rent. Conversely, if the land included within the demise is too small (for example, there is no room to expand or build a car park) then a lower rent (or initial premium) is likely to be paid.

- Payment. The rent (and any mechanism for review) will generally reflect the level of any initial premium paid for the site; if a large sum has been paid to secure the site, then minimal rents may be charged, with the site being developed on a ground lease rather than sold purely to enable the landlord to control the ongoing use of the site.

   Where the rent is based upon a percentage of turnover (typical at airports) we would expect to see clauses to enable the landlord to exercise some control over this income, whether by allowing them to review and comment upon the accounts or to enforce repairing obligations upon the tenant.

- The use. A number of ground leases will specifically state how the site can be used (and any building on them), enabling the landlord to have control of the use of the site throughout the term of the lease. A lease without this limitation on the use of the property is more valuable to the tenant.

- Repairs. Where the landlord is keen to control the ongoing use of the site as part of a larger "estate management" programme, then fixed repairing requirements may be set whether to ensure the usual "good repair" covenants or to insist that a proportion of turnover is spent on repairs.

Every ground lease varies and certain leases will have different "key clauses"; the valuer will need to review all the clauses to see what impact they would have on value.

Traditional building leases

A traditional building lease is typical of the property market as a whole. The hotel building will be leased to the operator and typically the term will be shorter and more restrictions will be placed than for a ground lease. Some of the key terms are as follows:

- Demise. Whether the property is fully fitted, built to a shell and core and where it extends to. In older leases, it is usual for hotel leases to be let on a shell and core basis, allowing the operator to fit out the hotel according to their own specific operational requirements.

- Term. The length of the lease can vary, typically from 15 years to 84 years, although shorter terms seem to be becoming more commonplace. Break clauses can be included within the terms of the lease, allowing one or either party to the lease to terminate the lease early. The desirability of the lease length will depend in the relative strengths of the market: in some markets, tenants will be prepared to pay a premium for a longer term, whereas in other market conditions break clauses may be more valuable to tenants.

- Rent reviews. How often the rent is reviewed will often have an impact upon the value of the investment (for both the landlord and the tenant). The methodology behind the review will also be important, with the additional certainty given by RPI-linked reviews prized by certain investors above potentially higher (if more risky) uplifts associated with review to the market rent (or proportion of capital value).

- Rent (and how it is calculated). There are many different ways to structure the rent, from straight turnover leases to part-fixed, part-EBITDA based rents, to fixed rents annually increased. The type (and level in relation to market norms and affordability) of this rent will affect the desirability of the investment.

    It is interesting to mention now that there can be a largely unforeseen problem with turnover leases. To illustrate this let us assume that a hotel with 400 bedrooms pays 25% of turnover by way of rent, which could be within market norms. In this hotel, there is a fine dining restaurant that runs at a profit margin of 15% and a quality spa that generates 20% profit. In this case, the tenant loses money for every pound that is generated in either of these two facilities, and it would therefore be in the tenant's financial interest to close down both operations. It is unlikely that it was the intention of either party when the lease was agreed to have both parties working at cross purposes (the essence of a turnover lease is that it is a partnership — in good times the operator can pay the landlord more rent while in poor times the landlord is prepared to accept less rent to help the tenant to work through the lean times). As such, it could be considered sensible to have differing turnover percentages for differing parts of the business, as long as where the revenue is allocated is clearly defined.

- Alienation. One of the key investment criteria for an investor is the quality of the tenant. As such, it is important that the clauses that allow for the tenant to change are reviewed and priced accordingly. Although absolute prohibitions on the tenant assigning their interest can be considered onerous to the tenant (potentially impacting on the length of lease they will consider and the rent they are prepared to pay), it is important that prohibitions ensure the quality of the operator can be dictated by the owner to ensure the investment value does not decline.

    Specific alienation provisions for hotels can include the proposed assignee operating more than a specific numbers of hotels or bedrooms (to ensure the experience of the assignee), specifying direct market experience appropriate to the hotel in question, specifying the financial standing of the proposed assignee is equal or better than the existing tenant, and specifying that the landlord's interest is not adversely affected by the assignment.

    In the case of turnover leases, there is often a blanket prohibition on subletting as this can adversely affect the landlord's income. For example, a hotelier may earn £500,000 revenue from a restaurant/bar in the hotel each year of which £125,000 converts to operating profit (25%). If a third-party tenant is prepared to pay £200,000 rent for the property it is in the financial interest of the operator to sublet the property. However, assuming the landlord is receiving 20% of turnover, in the case where the operator runs the restaurant themselves the landlord receives £100,000 rent from this part of the building, whereas when it was sublet out they only received £40,000.

- Repairs. Who pays for what repairs will have an impact on the value of the investment, with full repairing and insuring leases (where the tenant is responsible for the repairs to both the exterior and the interior of the property) still the standard within the hotel industry, although more and more internal repairing leases (where the tenant is only responsible for the interior of the building and not the external structure) are being seen, along with various hybrids of both types where part of the outside structure is maintained by the owner and the remaining repairing liability remains with the tenants.

- Insurance. It is usual for the tenant to bear the responsibility for the cost of insurance, although the landlord will be keen to ensure that they can have an adequate mechanism to insist that adequate insurance is in place. However, it can be more expensive for a hotel company to have the insurance purchased by the owner and then recharged to them, because some of the larger hotel companies are able to command significant cost discounts by placing a block insurance policy across their whole portfolio, rather than on an individual-by-individual property basis.

- Alterations. A blanket restriction on alterations can be prohibitive and will potentially effect the rent that a tenant will be prepared to pay, especially if the term of the lease is longer than 15 years, as the hotel industry requires flexibility to be able to ensure the property can match the needs of an evolving customer base.

- Keep open clauses. These are rarely used in hotel leases, although they are sometimes used in conjunction with turnover leases. Most hotels need to remain open to trade and so, unless the business is seasonal, it is not considered unduly onerous for a keep open clause to be included in the lease terms.

- User clauses. These can be quite simple (to use as a first-class hotel) but they can also have an impact upon the potential trading of the property (not to sell alcohol, not to hold auctions), as well as restricting other potentially more profitable alternative uses of the property.

- Optimise/maximise the turnover/EBITDA. In most leases where turnover or EBITDA are directly linked with the rent payable, there is often an obligation on the tenant to maximise (or optimise) trade.

- Asset management rights for the landlord. In most performance-based leases, there are the rights for the landlord to review the basic marketing plans and budgets of the tenant company to ensure that the company is doing its best to make as much money as possible (and therefore pay as much rent as possible). If the landlord does not have these rights it can adversely impact on the long-term value of the investment.

- Access to operational management accounts. Ideally, the lease should provide the landlord with access to the management accounts for the hotel. When the investment is being valued, revalued or sold, having access to the operational accounts will be beneficial to the investor, who will then have some certainty as to the affordability of the rent and the vacant possession value of the unit.

- Forfeiture. Clauses that are considered unduly onerous and affect on lending institutions' security over the asset will have an adverse impact on the value of the landlord's investment.

- Mandatory repairing (and furniture, fittings and equipment) expenditure. Some leases detail specific provisions for repair and maintenance, specifying an annual minimum spend. If this is set at the right level, it will be attractive to the landlord as it will ensure the property is kept in good condition whilst not being a disadvantage to the tenant. However, if the repairing spend is set too high it can be a hindrance to the tenant, who will unable to offer as high a rent as they otherwise would have done (and indeed it will deter some potential tenants), thereby affecting the value of the landlord's investment.

Management contracts

A management contract is similar to a turnover lease, but has the advantage of avoiding a balance sheet liability for the operator, while providing easier provisions for re-entry for an owner when the operator is not performing. However, there are three key differences between a turnover lease and a management contract: the employment liability and the repairing liability remain with the landlord, along with all the operational risks associated with the cost of running the business.

Some of the key terms in a management contract are as follows:

- Term. Some agreements provide that, at the operator's option, the relationship will last for up to 50 years, which can have an adverse effect upon the value of the hotel. Most terms tend to be for between 10 and 25 years and any period outside this range could influence the value of the investment. Whether extensions to the term are mutual or at the discretion of the operator can also have an impact on the value of the investment.

- Management fees. The cost of the management of the hotel by the operator will affect the cash flow generated by the owner, and as such will have an impact upon the value of the investment. It is usual to see a base fee (which is a percentage of turnover) and an incentive fee (usually a percentage of adjusted gross operating profit). The level of fees will vary depending on the locality but fees in the region of 2–3% of Revenue as a base fee and 8-10% of Gross Operating Profit as an incentive fee are typical. Certain contracts allow for incentive fees to be subordinated to debt service or some other agreed performance criteria. Subordinated fees may be waived or accrued, depending on the relative strength of the negotiating parties when the agreement was being drawn up. In certain circumstances, unpaid fees become debts, chargeable upon the property that fit the value of the investment.

- The operator system fees. The way that other charges are handled will affect the income generated by the investment and therefore its value. Reservations fees can be charged per reservation or per room night, or they can be a percentage of revenue generated or a fixed price per booking. They can be charged at the time of booking or when the guest arrives — sometimes reservation fees accrue in the event of no shows. All of these factors can have an influence on the value of the interest to the investor.

- Shared services. It is usual for the owner to have approval of those services that are to be shared with other hotels operated by the operator. These services should be detailed in the agreement. If there is no control over such shared facilities (for example the auditor for another (competitive) hotel run by the management company has the right to review the hotel accounts), it can have a detrimental impact upon the value of the investment.

- Performance criteria. Most management contracts contain no performance criteria at all, although the better-drafted contracts (from an investor's perspective) provide the right for the owner to terminate the agreement, without compensation, should the operator consistently fail to meet the budgeted targets, or underperform the competitive market. Choosing the competitive set for benchmarking purposes is quite subjective and these can sometimes be unsuitable, which can adversely affect the value of this clause to the investor.

- Owner's right to sell the property. The owner should be able to sell the hotel without the operator's consent, although many old agreements have a blanket restriction on the owner selling the hotel without the operator's approval. If a purchaser wishes to change management companies, or is itself a management company, there is usually a right to terminate the management agreement with compensation paid to the operating company.
  In some management contracts, the operator has a right of pre-emption and can take over the deal after it has been negotiated. This can have a dramatic impact on the value of the investment as it substantially reduces the number of investors who would be prepared to go through the due diligence process to purchase the property, knowing that at the end of the process it can have all be in vain, despite the truest intentions of the vendor, as they will be left with the very real expenses they have incurred through the process.

- Owner to approve budgets. Most management contracts state each year that the operator will provide detailed operating budgets, capital expenditure budgets and a marketing plan for the owner's approval. If the owner does not have this right, it could adversely affect the owner's investment.

- Owner to approve general manager and financial controller. The owner should have the right to approve the appointment of the hotel's general manager and other key personnel, such as the financial controller. Without the ability to question the appointment of the general manager the owner is in a weaker position to ensure the optimum performance of the investment.

- Replacement reserves. These should be appropriate to the quality of the asset, whether set at 3%, 4% per annum or at higher levels of turnover. It is in the owner's interest to have this clearly specified, because if an adequate furniture, fittings and equipment (FF&E) reserve is not deducted such expenditure will become a capital expense. Although there may be tax advantages to such a course of action it will result in higher incentive fees being paid to the operator. Any shortfall in FF&E expenditure should be carried over to the next year, and should not be credited to the gross operating profit (otherwise additional management fees may be paid on such underspend).

- Redevelopment if the property is destroyed. The owner should have the freedom to chose not to rebuild the hotel in the event that it is destroyed, as it can have a potentially higher alternative use value, which can be desirable to the investor.

- Restrictions on competition. The management contract may limit the number of hotels that the management company can operate within a "competitive" geographic locality to the subject hotel. It may be that the restriction is merely on the use of the brand name of the unit. The greater the restrictions the more control the investor has over the future openings affecting his investment. Alternatively (or in addition), the owner may have a priority listing in the operator's reservation system (ie, always first choice) or the right to approve any additional use of the operator's name.

- Bank accounts to be the property of the owner. The bank accounts are usually held in the name of the owner and not "held in trust" by the operator. The owner will usually also have the right to approve the signatories to the account. If they do not have these rights it can be detrimental to the value of the investment.

- Branded operator's equipment. This should be kept to a minimum in order to avoid any additional expense on behalf of the investor on termination of the management contract. If the level of branded equipment is unusually high, then it could have a damaging effect on the residual value of the investment.

- Arbitration. Most contracts allow for arbitration in the event of disputes between the parties. However, the chosen location of such arbitration and the law that is applicable can have quite an impact on the value of the investment.

- Independent external auditors. The owner should have the right to appoint his own accountants and auditors. If he does not have this right, potential investors may be deterred from taking on the investment for fear of their advisors not being seen to be sufficiently independent from the operator and thereby affecting the value.

Other methodologies for assessing the appropriate yield

Sometimes however a valuer is asked to assess the appropriate yield of a property where there is no established investment market and no comparable evidence. There are a number of factors that combine to determine the appropriate capitalisation rates that are used in the market and when no comparable evidence is available these should be looked at to help provide an opinion of value.

These factors can include the following:

- Cost of debt.
- Cost and availability of equity.
- Location risk.
- Property condition risk.
- Property category risk.
- Operational risk.
- Economic risk.
- Country risk.

This is a substantially more theoretical approach than using comparable evidence, and runs the risk of not necessarily reflecting the approach that would be taken by a potential purchaser. However, if done carefully, and if all the differing risk elements are accurately reflected, it can provide a decent starting point when no direct evidence is available.

However, where market evidence is available this should always be referred to ahead of calculating the capitalisation rate from scratch.

Another more suitable method for building up hotel investment yields without direct market evidence may be to look at other investments classes available and to adjust the returns available elsewhere to reflect the nature of the property. Alternatively, it may be appropriate to review similar transactions in similar geographic locations and then make relevant adjustments based on the differences in the area.

Example 21: Hotel Bridgetown

In this example, the hotel is a new development which will be operated on a management contract by an excellent international operator with no guaranteed income. It is in an area that has no existing international hotel brands, however, there are a number of comparable tenants in other classes of property.

It is known that offices let to similar quality tenants trade off yields of 8%, and retail units off 7.4%. The valuer will then have to make an adjustment based upon the uncertainty of the income stream, an adjustment due to the legal differences between a management contract and a lease and then an adjustment because of the type of property.

In this case, it is decided that the retail unit provides the best starting point and that an adjustment of 1.5% is needed to reflect the uncertainty of the income, 1% to reflect the additional liabilities to the investor of a management contract, no adjustment for the operator over the tenant, and 0.5% to reflect the different property class. This results in the valuer adopting a capitalisation rate of 10.4%.

Example 22: Hotel Scarborough

In this example, the hotel is a new development which will be operated on a management contract by an excellent international operator; there will be no guaranteed income. It is located in an area that has no existing international hotel brands and indeed few international quality tenants in any of the properties in the city, whether commercial retail or industrial.

The same management contract sells for 7.75% in the adjoining more stable country, which has a much more mature property market. It is considered likely that the difference in two countries is in the region of 2–3%, so the appropriate capitalisation rate is likely to be in the region of 9.75–10.75% for the property.

Problems with comparable evidence

Ideally, the valuer will be able to refer to market evidence to determine the appropriate multiple to apply to the EBITDA. The best evidence of what yield to use comes from recent market evidence of comparable properties, although there are problems with comparable evidence, which we outline below.

(i) Accuracy of the information provided and lack of transparency in the hotel transactional marketplace

The hotel property market is a relatively secretive world and transactional information is rarely available to the general public. Even when the price paid is accurately reported, it is highly unlikely that EBITDA numbers and trading projections will be fully available to accurately analyse the multiplier that was applied.

There is definitely a ranking system for comparable evidence in which first-hand detailed, transactional knowledge is the most desirable sort of evidence moving down the scale to partial hearsay evidence.

A valuer should be aware of the most comparable transactions in the market and hopefully will have a database that analyses the yields in transactions so that the valuer can be satisfied that his valuation is reflecting the current market.

It may be that only publicly reported facts can be included within the valuation report (to avoid breaching confidentiality agreements) but the valuer should draw the client's attention to such specialist knowledge so as to provide comfort that these secret transaction details have been taken into account.

Comparable evidence that the valuer does not have first-hand knowledge of needs to be verifiable in order to be acceptable at court and as such partial hearsay evidence can only ever be used as the broadest of all benchmarking tools, to set the tone of the market rather than provide direct evidence.

It can sometimes be essential to refer to hearsay evidence despite its low level of reliability, for example, if a similar property has been sold in the same time period. In these instances, it is sensible to try and speak with the agent or valuer involved to try and find out as much information as possible. It is likely that the valuer or agent will appreciate your position and will want to help as much as possible without breaking their client's confidentiality. However, care must always be taken as partial information can be quite misleading, as is shown in example 23.

Example 23: Hotel Cairo

From a superficial look at reported evidence, it may seem evident that if a property with 100 bedrooms that is generating an EBITDA of $1,000,000 and sold for $10,000,000, then it sold for $100,000 per bedroom off a 10 times multiple.

Purchase price per bedroom $\quad \dfrac{\$10,000,000}{100} = \$100,000$ per bedroom

Yield $\quad \dfrac{\$1,000,000}{\$10,000,000} = 10\%$

However, detailed knowledge of the transaction could show that there is a proposal to increase the bedroom count by 50 rooms and expand the meeting facilities at a cost of $2,000,000, which will lead to an increase in EBITDA to $1,400,000.

Purchase price per bedroom	$\dfrac{\$10,000,000 + \$2,000,000 \text{ (costs)}}{150}$	= $80,000 per room
Yield	$\dfrac{\$1,400,000}{(\$10,000,000 + \$2,000,000)}$	= 11.66%

In that case, both the price per bedroom and yield has been incorrectly analysed. It is probable that the valuer or agent involved in the transaction will be able to correct errors in the information you have been given regarding a transaction rather than provide you with the facts of the transaction.

(ii) Comparability of the transactions

As stated earlier in Chapter 6, there are grave difficulties with the use of comparable evidence in the hotel world since rarely are there two hotels of a similar enough nature and trading profile to warrant anything other than using the transaction as a benchmark rather than hard evidence.

In addition, there is seldom a good supply of comparable evidence to enable the valuer to identify the perfect comparable transaction. Most locations do not have large stocks of hotels and therefore any local transaction tends to be looked at as a potential comparable property.

It is also common for valuers to look through comparable evidence further afield if the property type is comparable, for example a four-star hotel in Helsinki may sometimes be referred to by a valuer looking for a comparables for a similar hotel in Tallin, even though the dynamics of the two markets are different.

It is therefore normal for the comparable evidence to be adjusted (through the valuer's market knowledge and experience) to make it comparable to the subject property.

(iii) Analysis of the comparable evidence and the treatment of costs

It is important for the valuer to know what they are looking at when they are consulting comparable evidence. For example, the press may report that a particular hotel sold off an 8% yield and the valuer will need to analyse the details behind the transaction to see if it was correct.

Many times the reported earnings may not be accurately reported but, assuming that the EBITDA is correctly reported, the valuer must look to see how accurate it is, for example whether management costs and an FF&E reserve have been accurately deducted.

Example 24: Hotel Berlin

The press has reported that a 100-bedroom hotel sold for $10,000,000 (including costs) off an EBITDA of $1,000,000, suggesting a 10% capitalisation rate ($1,000,000/$10,000,000).

Purchase price per bedroom	$\dfrac{\$10,000,000}{100}$	= $100,000 per bedroom
Yield	$\dfrac{\$1,000,000}{\$10,000,000}$	= 10% capitalisation rate

It was subsequently discovered that the turnover was $3,000,000 and that a management fee and FF&E reserve had not been deducted when calculating the reported EBITDA. It is deemed appropriate that a 3% management fee is deducted along with a 4% FF&E reserve for this type of property.

Turnover	$3,000,000
Reported EBITDA	$1,000,000
Less FF&E reserve at 4%	$120,000
Less management fee at 3%	$90,000
Adjusted EBITDA	$790,000
Sales price (including costs)	$10,000,000
Yield	($790,000/$10,000,000) = 7.9%

It is important to say here that it is not always appropriate to deduct management costs or FF&E reserves when carrying out a valuation — the duty of the valuer is to reflect what the market does when assessing the value of that type of property. If it is inappropriate to deduct a management fee or an FF&E reserve, then the valuer should ensure that the comparables they refer to have been calculated on a similar basis.

Summary

Determining the appropriate yield to apply in a hotel valuation is an extremely complex business and requires both experience and full exposure to the current market.

All relevant comparable transactions must be fully analysed but, because of the individual nature of the business, it is likely that many adjustments will be required (or at the very least considered) to enable the value to correctly reflect the market's perception of the appropriate yield.

The Treatment of Capital Expenditure

What is capital expenditure?

In simple terms, capital expenditure (capex) is the amount of money that is required to improve the property that should not be written off, from an accounting perspective, over one year, as that would adversely affect the profitability of the property.

The Accounting Standards body defines capital expenditure as:

> Funds used by a company to acquire or upgrade physical assets such as property, industrial buildings or equipment. This type of outlay is made by companies to maintain or increase the scope of their operation. These expenditures can include everything from repairing a roof to building a brand new factory.

In terms of accounting, an expense is considered to be a capital expense when the asset is newly purchased or is an investment that improves the useful life of an existing capital asset. If an expense is a capital expense, it needs to be capitalised and depreciated; this requires the company to spread the cost of the expenditure over the useful life of the asset. If, however, the expense is one that maintains the asset at its current condition, the cost is deducted fully in the year of the outlay.

For example, if the cost of refurbishing a hotel is $2,500,000 and if it were deducted from the profit and loss account in one year, it would adversely impact the accounts and the hotel would be unduly unrepresentative of the actual trading of the property, with a large deficit in the accounts in one year.

However, in hotel terms capex can be split into two quite separate subsections: "defensive" capex and "offensive" capex. Defensive capex is money that is needed to be spent to maintain the current trading performance of a property, whereas offensive capex is money that is being spent to improve the trading potential of a property. "Defensive" capex should, in theory, be funded from the furniture, fittings and equipment (FF&E) reserve assuming that the level of provision in the FF&E reserve is adequate. It is not always clear which is which, for example, a bedroom refurbishment is likely to be part defensive and part offensive capex. They are both treated the same, however, from an accounting perspective.

In terms of valuation, it is more usual to include defensive capex into the valuation rather than offensive capex, although when there is a clear "offensive" option that most prospective "reasonably efficient operators" would undertake, then offensive capex must be taken into consideration in the valuation process.

Table 8.1 Forecast in present values, Hotel Lisbon

Year	1		2		3	
No of Rooms	100		100		100	
Rooms Sold	25,550		25,550		25,550	
Rooms Available	36,500		36,500		36,500	
Occupancy (%)	70.0		70.0		70.0	
ADR	85.00		85.00		85.00	
RevPAR	59.50		59.50		59.50	
Growth (RevPAR) (%)			0.0		0.0	
Revenues (€000s)		(%)		(%)		(%)
Rooms	2,171.8	82.0	2,171.8	82.0	2,171.8	82.0
Food	132.4	5.0	132.4	5.0	132.4	5.0
Beverage	185.4	7.0	185.4	7.0	185.4	7.0
Total Food & Beverage	317.8	12.0	317.8	12.0	317.8	12.0
Room Hire	0.0	0.0	0.0	0.0	0.0	0.0
Leisure Club	105.9	4.0	105.9	4.0	105.9	4.0
Other Income	53.0	2.0	53.0	2.0	53.0	2.0
Total Revenue	**2,648.5**	**100.0**	**2,648.5**	**100.0**	**2,648.5**	**100.0**
Departmental Profit						
Rooms	1,411.6	65.0	1,411.6	65.0	1,411.6	65.0
Total Food & Beverage	111.2	35.0	111.2	35.0	111.2	35.0
Room Hire	0.0	100.0	0.0	100.0	0.0	100.0
Leisure Club	90.0	85.0	90.0	85.0	90.0	85.0
Other Income	18.5	35.0	18.5	35.0	18.5	35.0
Total Departmental Profit	1,631.5	61.6	1,631.5	61.6	1,631.5	61.6
Departmental Costs	1,017.0	38.4	1,017.0	38.4	1,017.0	38.4
Undistributed Operating Expenses						
Administrative & General	238.4	9.0	238.4	9.0	238.4	9.0
Sales & Marketing	132.4	5.0	132.4	5.0	132.4	5.0
Property Operations & Maintenance	105.9	4.0	105.9	4.0	105.9	4.0
Utility Costs	92.7	3.5	92.7	3.5	92.7	3.5
Total Undistributed Expenses	569.4	21.5	69.4	21.5	569.4	21.5
Income Before Fixed Costs	**1,062.0**	**40.1**	**1,062.0**	**40.1**	**1,062.0**	**40.1**
Fixed Costs						
Reserve for Renewals	105.9	4.0	105.9	4.0	105.9	4.0
Property Taxes	84.8	3.2	84.8	3.2	84.8	3.2
Insurance	29.1	1.1	29.1	1.1	29.1	1.1
Management Fees	79.5	3.0	79.5	3.0	79.5	3.0
Total Fixed Costs	299.3	11.3	299.3	11.3	299.3	11.3
EBITDA	**762.8**	**28.8**	**762.8**	**28.8**	**762.8**	**28.8**

Source: Leisure Property Services

It is important that the valuer ensures that any capex that has been incorrectly deducted from the profit and loss (usually under the Repairs & Maintenance section) is added back when calculating the EBITDA, and then deducted from the valuation as shown in example 25. If this is not done, the valuation will be mathematically incorrect unless appropriate adjustments are made to the FF&E reserve and the capitalisation rate.

To undertake this, it is necessary to analyse the detail behind any allowances made for the FF&E reserve and under the Repairs & Maintenance department to ensure all costs are adequately allowed for.

Example 25: Hotel Lisbon

In this example, the hotel is required to spend £500,000 to ensure that trading will remain constant at the projected levels, and that this work can be carried out without affecting the trading of the hotel.

The above "shadow" projections show that the proposed capex will not affect the trading either positively or negatively. (Note the projections are in present values, meaning inflation has not been included in the cash flows).

Table 8.2 Income Capitalisation, Hotel Lisbon

Net Base Cash Flow		
EBITDA in 1 values in year 1	762,761	
Capitalised at 10.00%	10.00	7,627,610
Less		
Capital Expenditure		500,000
Gross Value		7,127,610
Say		7,125,000

Source: Leisure Property Services

As can be seen, the value of the property is lower, having been diminished by the cost of the capex, and results in a value of £7,125,000.

Example 26: Hotel Oslo

In this example, it is proposed to extend the property by 10 bedrooms, with an extension being built to the side of the existing building. The cost of the works will be €650,000 and the work will be completed in 12 months. The new bedrooms are expected to be open in the second year of trading.

It is anticipated that trading in the second year will experience a temporary dip in volume of business but this will stabilise the year after. The location of the new building works means that the current facilities will not be affected.

It should be noted in this example that the additional bedrooms were absorbed into the trading inventory in two years, trading at the same stabilised occupancy and ADR as the smaller hotel. Also, the 10% increase in size did not affect the profit margin of the hotel, with the extra 10% capacity resulting in incremental increases in departmental costs, undistributed costs and fixed costs.

Table 8.3 Forecast in present values, Hotel Oslo

Year	1		2		3	
No of Rooms	100		110		110	
Rooms Sold	25,550		27,302		28,105	
Rooms Available	36,500		40,150		40,150	
Occupancy (%)	70.0		68.0		70.0	
ADR	85.0		85.0		85.0	
RevPAR	59.5		57.8		59.5	
Growth (RevPAR) (%)			−2.9		2.9	
Revenues (€000s)		(%)		(%)		(%)
Rooms	2,171.8	82.0	2,320.7	82.0	2,388.9	82.0
Food	132.4	5.0	141.5	5.0	145.7	5.0
Beverage	185.4	7.0	198.1	7.0	203.9	7.0
Total Food & Beverage	317.8	12.0	339.6	12.0	349.6	12.0
Leisure Club	105.9	4.0	113.2	4.0	116.5	4.0
Other Income	53.0	2.0	56.6	2.0	58.3	2.0
Total Revenue	**2,648.5**	**100.0**	**2,830.1**	**100.0**	**2,913.3**	**100.0**
Departmental Profit						
Rooms	1,411.6	65.0	1,508.4	65.0	1,552.8	65.0
Total Food & Beverage	111.2	35.0	118.9	35.0	122.4	35.0
Leisure Club	90.0	85.0	96.2	85.0	99.1	85.0
Other Income	18.5	35.0	19.8	35.0	20.4	35.0
Total Departmental Profit	**1,631.5**	**61.6**	**1,743.3**	**61.6**	**1,794.6**	**61.6**
Departmental Costs	**1,017.0**	**38.4**	**1,086.8**	**38.4**	**1,118.7**	**38.4**
Undistributed Operating Expenses						
Administrative & General	238.4	9.0	254.7	9.0	262.2	9.0
Sales & Marketing	132.4	5.0	141.5	5.0	145.7	5.0
Property Operations & Maintenance	105.9	4.0	113.2	4.0	116.5	4.0
Utility Costs	92.7	3.5	99.1	3.5	102.0	3.5
Total Undistributed Expenses	**569.4**	**21.5**	**608.5**	**21.5**	**626.4**	**21.5**
Income Before Fixed Costs	**1,062.0**	**40.1**	**1,134.9**	**40.1**	**1,168.2**	**40.1**
Fixed Costs						
Reserve for Renewals	105.9	4.0	113.2	4.0	116.5	4.0
Property Taxes	84.8	3.2	90.6	3.2	93.2	3.2
Insurance	29.1	1.1	31.1	1.1	32.0	1.1
Management Fees	79.5	3.0	84.9	3.0	87.4	3.0
Total Fixed Costs	**299.3**	**11.3**	**319.8**	**11.3**	**329.2**	**11.3**
EBITDA	**762.8**	**28.8**	**815.1**	**28.8**	**839.0**	**28.8**

Source: Leisure Property Services

The Treatment of Capital Expenditure

Table 8.4 Income capitalisation, Hotel Oslo

Net Base Cash Flow			
EBITDA in 1 values in year 3		839,037	
Capitalised at 10.00%		10.00	8,390,371
Less	Income Shortfall		100,249
	Capital Expenditure		650,000
	Subtotal		750,249
Gross Value			7,640,122
Say			7,640,000

Source: Leisure Property Services

It can be seen that the ability to add on 10 extra bedrooms at a cost of €65,000 per bedroom has increased the value of the overall property at the current time. The value will further increase when the capital expenditure has been spent.

Example 27: Hotel Zagreb

Sometimes the capital expenditure programme will have a direct impact on the trading of the hotel while the works are being undertaken.

In this example, the 150-bedroom hotel will have all of the bedrooms refurbished over the first six months of the year, with 25 bedrooms being taken off the letting inventory for four weeks at a time. During the same period, the meeting rooms will be completely refurbished and the restaurant and bar will be repainted and provided with new furniture and carpets.

The total cost of the works will be £2,000,000 and all of the work is expected to be completed in six months.

Table 8.5 Income capitalisation, Hotel Zagreb

Net Base Cash Flow			
EBITDA in 1 values in year 3		2,460,278	
Capitalised at 8.25%		12.12	29,821,552
Less	Income Shortfall		1,699,654
	Capital Expenditure		2,000,000
	Subtotal		3,699,654
Gross Value			26,121,898
Say			26,120,000
Price per Bedroom			174,133

Source: Leisure Property Services

Table 8.6 Forecast in present values, Hotel Zagreb

Year	1		2		3	
No of Rooms	150		150		150	
Rooms Sold	35,040		37,230		38,325	
Rooms Available	54,750		54,750		54,750	
Occupancy (%)	64.0		68.0		70.0	
ADR	135.00		140.00		160.00	
RevPAR	86.40		95.20		112.00	
Growth (RevPAR) (%)			10.2		17.6	
Revenues (£'000s)		(%)		(%)		(%)
Rooms	4,730.4	82.0	5,212.2	82.0	6,132.0	82.0
Food	288.4	5.0	317.8	5.0	373.9	5.0
Beverage	403.8	7.0	444.9	7.0	523.5	7.0
Total Food & Beverage	692.3	12.0	762.8	12.0	897.4	12.0
Leisure Club	230.8	4.0	54.3	4.0	299.1	4.0
Other Income	115.4	2.0	127.1	2.0	149.6	2.0
Total Revenue	**5,768.8**	**100.0**	**6,356.3**	**100.0**	**7,478.0**	**100.0**
Departmental Profit						
Rooms	2,838.2	60.0	3,387.9	65.0	4,292.4	70.0
Total Food & Beverage	207.7	30.0	267.0	35.0	314.1	35.0
Leisure Club	196.1	85.0	216.1	85.0	254.3	85.0
Other Income	40.4	35.0	44.5	35.0	52.3	35.0
Total Departmental Profit	**3,282.4**	**56.9**	**3,915.5**	**61.6**	**4,913.1**	**65.7**
Departmental Costs	**2,486.3**	**43.1**	**2,440.8**	**38.4**	**2,565.0**	**34.3**
Undistributed Operating Expenses						
Administrative & General	519.2	9.0	572.1	9.0	673.0	9.0
Sales & Marketing	288.4	5.0	317.8	5.0	373.9	5.0
Property Operations & Maintenance	230.8	4.0	254.3	4.0	299.1	4.0
Utility Costs	201.9	3.5	222.5	3.5	261.7	3.5
Total Undistributed Expenses	**1,240.3**	**21.5**	**1,366.6**	**21.5**	**1,607.8**	**21.5**
Income Before Fixed Costs	**2,042.1**	**35.4**	**2,548.9**	**40.1**	**3,305.3**	**44.2**
Fixed Costs						
Reserve for Renewals	230.8	4.0	254.3	4.0	299.1	4.0
Property Taxes	184.6	3.2	203.4	3.2	239.3	3.2
Insurance	63.5	1.1	69.9	1.1	82.3	1.1
Management Fees	173.1	3.0	190.7	3.0	224.3	3.0
Total Fixed Costs	**651.9**	**11.3**	**718.3**	**11.3**	**845.0**	**11.3**
EBITDA	**1,390.3**	**24.1**	**1,830.6**	**28.8**	**2,460.3**	**32.9**

Source: Leisure Property Services

The Treatment of Capital Expenditure

It is clear that the loss of the bedrooms throughout the refurbishment has adversely affected the hotel's trading. The volume of business decreased and the rate achievable also fell.

This is quite common as it is difficult for the upgrade of an operating hotel to be carried out without any impact on the trading. The hotel is less likely to attract meetings or events if noisy construction work is going on during the day; weddings and other events are also likely to be discouraged by the hotel being physically "unfinished".

Timing of capital expenditure

The timing of the capex is important from a valuation perspective both in terms of when the money has to be actually spent and in terms of when the improved product is available for trading This is illustrated by the two examples below.

Example 28: Hotel Cardiff

In example 28, Hotel Cardiff a 100-bedroom hotel, is spending €400,000 in defensive capex to maintain the market position of the hotel in one year, with the rooms being repainted and new carpets being fitted. It is not anticipated that trading will be effected by the works being carried out.

Table 8.7 Income capitalisation, Hotel Cardiff

Net Base Cash Flow			
EBITDA in 1 values in year 1		987,102	
Capitalised at 10.00%		10.00	9,871,024
Less	Income Shortfall		–
	Capital Expenditure		400,000
	Subtotal		400,000
Gross Value			9,471,024
Say			9,470,000

Source: Leisure Property Services

Example 29: Hotel Beijing

In example 29 Hotel Beijing has exactly the same trading profile and is carrying out the same works, but this time over four years. The timing of the work has been set so that the worst bedrooms will be upgraded immediately with the better condition rooms being left until last, with the effect that the trading will remain constant.

Valuation of Hotels for Investors

Table 8.9 Forecast in present values, Hotel Cardiff and Hotel Beijing

Year	1		2		3	
No of Rooms	100		100		100	
Rooms Sold	25,550		25,550		25,550	
Rooms Available	36,500		36,500		36,500	
Occupancy (%)	70.0		70.0		70.0	
ADR	110.00		110.00		110.00	
RevPAR	77.00		77.00		77.00	
Growth (RevPAR) (%)			0.0		0.0	
Revenues (€'000s)		(%)		(%)		(%)
Rooms	2,810.5	82.0	2,810.5	82.0	2,810.5	82.0
Food	171.4	5.0	171.4	5.0	171.4	5.0
Beverage	239.9	7.0	239.9	7.0	239.9	7.0
Total Food & Beverage	411.3	12.0	411.3	12.0	411.3	12.0
Room Hire	0.0	0.0	0.0	0.0	0.0	0.0
Leisure Club	137.1	4.0	137.1	4.0	137.1	4.0
Other Income	68.5	2.0	68.5	2.0	68.5	2.0
Total Revenue	**3,427.4**	**100.0**	**3,427.4**	**100.0**	**3,427.4**	**100.0**
Departmental Profit						
Rooms	1,826.8	65.0	1,826.8	65.0	1,826.8	65.0
Total Food & Beverage	144.0	35.0	144.0	35.0	144.0	35.0
Room Hire	0.0	100.0	0.0	100.0	0.0	100.0
Leisure Club	116.5	85.0	116.5	85.0	116.5	85.0
Other Income	24.0	35.0	24.0	35.0	24.0	35.0
Total Departmental Profit	**2,111.3**	**61.6**	**2,111.3**	**61.6**	**2,111.3**	**61.6**
Departmental Costs	**1,316.1**	**38.4**	**1,316.1**	**38.4**	**1,316.1**	**38.4**
Undistributed Operating Expenses						
Administrative & General	308.5	9.0	308.5	9.0	308.5	9.0
Sales & Marketing	171.4	5.0	171.4	5.0	171.4	5.0
Property Operations & Maintenance	137.1	4.0	137.1	4.0	137.1	4.0
Utility Costs	120.0	3.5	120.0	3.5	120.0	3.5
Total Undistributed Expenses	**736.9**	**21.5**	**736.9**	**21.5**	**736.9**	**21.5**
Income Before Fixed Costs	**1,374.4**	**40.1**	**1,374.4**	**40.1**	**1,374.4**	**40.1**
Fixed Costs						
Reserve for Renewals	137.1	4.0	137.1	4.0	137.1	4.0
Property Taxes	109.7	3.2	109.7	3.2	109.7	3.2
Insurance	37.7	1.1	37.7	1.1	37.7	1.1
Management Fees	102.8	3.0	102.8	3.0	102.8	3.0
Total Fixed Costs	**387.3**	**11.3**	**387.3**	**11.3**	**387.3**	**11.3**
Incentive Management Fee	0.0	0.0	0.0	0.0	0.0	0.0
EBITDA	**987.1**	**28.8**	**987.1**	**28.8**	**987.1**	**28.8**

Source: Leisure Property Services

Table 8.9 Income capitalisation, Hotel Beijing

Net Base Cash Flow			
EBITDA in 1 values in year 1		987,102	
Capitalised at 10.00%		10.00	9,871,024
Less	Income Shortfall		–
	Capital Expenditure		385,602
	Subtotal		385,602
Gross Value			9,485,422
Say			9,485,000

Source: Leisure Property Services

It is clear that the net present value of the capital expenditure is lower if it is spread over the four years, leading to an increase in the value of the hotel.

Example 30: Hotel Budapest

However, in example 30 Hotel Budapest, the same delay in improving the bedrooms impacts upon the trading of the "unimproved rooms" leading to deterioration in the trading and value as shown in Table 8.11.

Table 8.10 Income capitalisation, Hotel Budapest

Net Base Cash Flow			
EBITDA in 1 values in year 5			987,102
Capitalised at 10.00%		10.00	9,871,024
Less	Income Shortfall		263,761
	Capital Expenditure		385,602
	Subtotal		649,364
Gross Value			9,221,661
Say			9,220,000
Price per Bedroom			92,200

Source: Leisure Property Services

In this instance, even though the net present values of the capital expenditure is lower than spending the money straight away, the adverse impact on trade until the property has been upgraded results in a value that is €250,000 lower.

Table 8.11 Forecast in present values, Hotel Budapest

Year	1		2		3		4	
No of Rooms	100		100		100		100	
Rooms Sold	24,820		24,455		24,455		25,550	
Rooms Available	36,500		36,500		36,500		36,500	
Occupancy (%)	68.0		67.0		67.0		70.0	
ADR	105.00		104.50		103.00		110.00	
RevPAR	71.40		70.02		69.01		77.00	
Growth (RevPAR) (%)			−1.9		−1.4		11.6	
Revenues (€000s)		%		%		%		%
Rooms	2,606.1	82.0	2,555.5	82.0	2,518.9	82.0	2,810.5	82.0
Food	158.9	5.0	155.8	5.0	53.6	5.0	171.4	5.0
Beverage	222.5	7.0	218.2	7.0	215.0	7.0	239.9	7.0
Total Food & Beverage	381.4	12.0	374.0	12.0	368.6	12.0	411.3	12.0
Leisure Club	127.1	4.0	124.7	4.0	122.9	4.0	137.1	4.0
Other Income	63.6	2.0	62.3	2.0	61.4	2.0	68.5	2.0
Total Revenue	3,178.2	100.0	3,116.5	100.0	3,071.8	100.0	3,427.4	100.0
Departmental Profit								
Rooms	1,694.0	65.0	1,661.1	65.0	1,637.3	65.0	1,826.8	65.0
Total Food & Beverage	133.5	35.0	130.9	35.0	129.0	35.0	144.0	35.0
Leisure Club	108.1	85.0	106.0	85.0	104.4	85.0	116.5	85.0
Other Income	22.2	35.0	21.8	35.0	21.5	35.0	24.0	35.0
Total Departmental Profit	1,957.8	61.6	1,919.8	61.6	1,892.2	61.6	2,111.3	61.6
Departmental Costs	1,220.4	38.4	1,196.7	38.4	1,179.6	38.4	1,316.1	38.4
Undistributed Operating Expenses								
Administrative & General	286.0	9.0	280.5	9.0	276.5	9.0	308.5	9.0
Sales & Marketing	158.9	5.0	155.8	5.0	153.6	5.0	171.4	5.0
Property Operations & Maintenance	127.1	4.0	124.7	4.0	122.9	4.0	137.1	4.0
Utility Costs	111.2	3.5	109.1	3.5	107.5	3.5	120.0	3.5
Total Undistributed Expenses	83.3	21.5	670.1	21.5	660.4	21.5	736.9	21.5
Income Before Fixed Costs	1,274.4	40.1	1,249.7	40.1	1,231.8	40.1	1,374.4	40.1
Fixed Costs								
Reserve for Renewals	127.1	4.0	124.7	4.0	122.9	4.0	137.1	4.0
Property Taxes	101.7	3.2	99.7	3.2	98.3	3.2	109.7	3.2
Insurance	35.0	1.1	34.3	1.1	33.8	1.1	37.7	1.1
Management Fees	95.3	3.0	93.5	3.0	92.2	3.0	102.8	3.0
Total Fixed Costs	359.1	11.3	352.2	11.3	347.1	11.3	387.3	11.3
EBITDA	**915.3**	**28.8**	**897.6**	**28.8**	**884.7**	**28.8**	**987.1**	**28.8**

Source: Leisure Property Services

Summary

Capital expenditure is an important consideration for the purchaser and valuer as the requirement to spend money will directly affect the value of the asset. Opportunities to enhance the trading potential of the asset should be reflected if they represent market sentiments, that is, if potential purchasers would look to make such an investment and would factor it into the acquisition appraisal.

Methodology for Assessing Market Rent

Introduction

In this chapter, we outline the methodology for assessing for the market rent of a hotel, which is in effect the annual value of a property (as opposed to the capital value of a property as discussed in Chapter 6).

Market rent is defined by the *Valuation & Appraisal Manual* (*Red Book*) as:

> The estimated amount for which a property, or space within a property, should lease on the date of valuation between a willing lessor and a willing lessee on appropriate lease terms, in an arm's length transaction, after proper marketing wherein the parties had each acted knowledgeably, prudently and without compulsion.

As such, we are trying to determine what potential tenants would be prepared to pay in the current marketplace, and therefore it is essential that the valuer reflects the approach adopted by that same market when deciding which methodology to adopt to calculate the market rent.

In most instances, the agreed rent on a property is almost exclusively determined by the potential profitability of the hotel/site albeit that the potential level of trade will be assessed using market evidence. It is not solely determined by comparable evidence is common with other property classes, although market comparables are readily referred to to see whether the end result appears reasonable.

The terms of the lease could have a direct effect on the profitability of the hotel, or they may influence the desirability of the property from an operational perspective. As such, the terms of the lease will have a key impact on the assessment of market rent.

The impact of lease terms on rental values

The wording and terms of the lease will affect the rental value of the property. One of the most important terms will be the rent review clause and the assumptions contained within it, including the following:

The assumed term

Is the assumed term longer or shorter than the market average, and would this impact on the rent that a hypothetical tenant would be prepared to accept?

When the length of "notional term" in the rent review clause is different to terms commonly accepted in the market, this could have an effect on the rental value at the review date. An assumed term at rent review of 40 years without a break, when the norm in the open market is for a term closer to 20 years could lead to an argument for a discount in the revised rent.

Conversely however, if the assumed term is too short to allow a tenant a sufficient period to write off fitting out expenses and marketing set-up expenses (in the case of a non-fully fitted property), the rental value could be reduced.

When determining whether the assured term is too long or too short the size of the hotel can sometimes be a relevant factor. For example, tenants leasing a hotel with 100+ bedrooms will be unlikely to accept a shorter term unless the rent is lower, whereas a tenant may be prepared to let a smaller hotel with 20 bedrooms on a short-term basis.

The effects of unusually shortened terms were to some extent mitigated by *Pivot Properties Ltd v Secretary of State for the Environment* (1980). This case established that the valuer must take into account the possibility of renewal under Part II of the Landlord and Tenant Act 1954. However, it is the duty of the valuer to reflect the open market; if the market is likely to worry about any of the section 31 grounds for possession being enforced, particularly the use by the owner or redevelopment, then the rent will be discounted by the shorter term.

The section 31 grounds are as follows:

(a) where under the current tenancy the tenant has any obligations as respects the repair and maintenance of the holding, that the tenant ought not to be granted a new tenancy in view of the state of repair of the holding, being a state resulting from the tenant's failure to comply with the said obligations;

(b) that the tenant ought not to be granted a new tenancy in view of his persistent delay in paying rent which has become due;

(c) that the tenant ought not to be granted a new tenancy in view of other substantial breaches by him of his obligations under the current tenancy, or for any other reason connected with the tenant's use or management of the holding;

(d) that the landlord has offered and is willing to provide or secure the provision of alternative accommodation for the tenant, that the terms on which the alternative accommodation is available are reasonable having regard to the terms of the current tenancy and to all other relevant circumstances, and that the accommodation and the time at which it will be available are suitable for the tenant's requirements (including the requirement to preserve goodwill) having regard to the nature and class of his business and to the situation and extent of, and facilities afforded by, the holding;

(e) where the current tenancy was created by the sub-letting of part only of the property comprised in a superior tenancy and the landlord is the owner of an interest in reversion expectant on the termination of that superior tenancy, that the aggregate of the rents reasonably obtainable on separate lettings of the holding and the remainder of that property would be substantially less than the rent reasonably obtainable on a letting of that property as a whole, that on the termination of the current tenancy the landlord requires possession of the holding for the purpose of letting or otherwise disposing of the said property as a whole, and that in view thereof the tenant ought not to be granted a new tenancy;

(f) that on the termination of the current tenancy the landlord intends to demolish or reconstruct the premises comprised in the holding or a substantial part of those premises or to carry out substantial work of construction on the holding or part thereof and that he could not reasonably do so without obtaining possession of the holding;

(g) subject as hereinafter provided, that on the termination of the current tenancy the landlord intends to occupy the holding for the purposes, or partly for the purposes, of a business to be carried on by him therein, or as his residence.

Of particular interest to hotel valuers is the repossession for their own use of the property by landlords. It should be noted that legally it is considered to be taking in a property for one's own use (ground g) if the landlord intends to let a manager operate the property in their behalf on a management contract.

That the tenant has complied with covenants

It is also usual for the rent review clause to assume that the tenant has complied with their covenants under the lease. Even if there is no specific assumption that the tenant has complied with their covenants under the lease, it would be unfair for the rent to be depressed as a result of a lack of repair or other breach of the tenant's covenants. *Harmsworth Pension Fund Trustees Ltd v Charrington Industrial Holdings Ltd* (1985) makes it clear the assumption that the tenant has complied with their covenants will be implied if not set out expressly.

Repair assumptions:

There are a number of considerations that must be taken into account in relation to the assumed condition of the premises for the purposes of review.

If the hotel is let on full repairing terms, then the rent review clause dictates assumption of compliance by tenant with covenants.

Where the review clause requires the valuer to assume that all the tenant's covenants have been complied with, no valuation problems should arise. Where the lessee is responsible for all repairs, the property will be valued as if in full repair.

In rare cases, there may be an issue as to whether defects existing at a review date are matters of repair covered by the repairing covenant or would require replacement, which would put them outside the scope of the repairing obligation, as shown in *Ravenseft Properties Ltd v Davstone (Holdings) Ltd* [1980].

The existence of a defect for which the tenant is not responsible could have very serious valuation consequences, as shown in *London Borough of Camden v Civil Aviation Authority Civil Aviation Authority* [1980]), a rating case where defects were held to make the property virtually unlettable.

If the landlord has covenanted to repair and the property is in disrepair, it is a matter of judgment as to whether the right for the tenant to enforce the landlord's covenant would wholly or partially offset any diminution in value attributable to the disrepair. In practice, any material disrepair is likely to adversely affect the rental value.

If the rent review clause stipulates that it shall be assumed that both landlord and tenant have complied with their repairing covenants, the valuer must value the property as if in good repair, even if this produces an injustice to the tenant, where the property is in disrepair through the landlord's failure to perform his covenant.

Alterations and improvements

It is usual for the lease to disregard any improvements made by the tenant. However, this can have a dramatic effect on the value where improvements, rather normal maintenance, have been made.

It is essential that the valuer knows exactly what they are valuing when trying to assess the rental valuation of a property. Questions such as: "Have any tenant's improvements been carried out that are excluded from the valuation, what is the specification of the building to be valued or does the demise of the leases include the furniture?" should be foremost in the valuer's mind.

Below we outline in detail the comparable method of valuation and the profits method of valuation and then highlight other lesser-used methodologies.

Comparable method

Rarely are there two hotels, in similar locations, with similar facilities that are trading in a similar enough manner to enable the rent offered in one hotel to be comparable enough to do more than just set a general tone for the letting negotiations.

All the problems that were outlined in Chapter 6 with regard to comparable evidence for capital valuation also apply to rental valuations.

When looking at comparable rental transactions, it is usual to adjust these as appropriate to reflect the prevalent trading conditions in the locality, the relative size and facilities of the hotel, as well as differences in dates, lease terms and underlying demand in the market.

Example 31: Hotel Ankara

In example 31, Hotel Ankara is one of a number of terraced properties on the seafront catering to the local domestic bed and breakfast leisure market. Each property is arranged on lower ground, ground and four upper floors and has between 18 and 20 guest rooms. All the hotels have the same landlord, and the lease terms are identical. The following new lettings and rent reviews shown in Table 9.1 have happened in the past three years.

Table 9.1 New lettings and rent reviews

Hotel	Beds	Rent (£)	Rent/bedroom (£)	Date	Type
Hotel Jersey	18	55,800	3,100	2 months ago	New letting
Hotel Guernsey	18	54,900	3,050	1 year ago	Rent review
Hotel Skye	20	62,000	3,100	1 year ago	New letting
Hotel Gozo	20	60,000	3,000	2 years ago	Rent review
Hotel Comina	20	63,000	3,150	2 years ago	Rent review
Hotel Ankara	20				Rent review

Source: Leisure Property Services

In this example, it is clear that all the evidence points to a rental value for the building of £62,000, or £3,100 per bedroom, and so the valuer is likely to assess the rental value for the property at £62,000 per annum.

Rental evidence from the analysis of comparable lettings will usually be the most highly prized evidence. It is important, however, for valuers to attach the relevant weight and importance to all the comparable evidence. The greatest weight will be given to transactions relating to properties as similar as possible to the subject property, and where the date of the transaction is as close as possible to the valuation date for the rent review in dispute.

Valuers may also be aware of transactions occurring after the date of valuation (in the event of a rent review). Such transactions are not primary (or genuinely relevant) evidence, since they could not have been known to the parties on the review date. They can, however, be useful to support the valuer's opinion as to movement in the marketplace.

The valuer will also give greater weight to different types of transaction. For example, new lettings are considered more valuable than rent review evidence, which can become self-generating, with errors being compounded.

The quality of the evidence derived from different sources will depend upon the circumstances of each case. The descending order of weight is:

1. Open market lettings
   An open market letting is the most important guide to the figure sought under most rent review provisions, as such provisions generally attempt to simulate what is taking place in the market without the property actually being placed on the market. As such, open market lettings carry substantially greater weight than the other evidence categories.

2. Agreements between valuers on lease renewal or rent review
   An agreement between valuers (for example, rent reviews or lease renewals) will usually reflect most of the evidence in the market at the relevant date, and it can be presumed that those valuers are likely to have reached a fair conclusion.

   It is important, however, to be aware of and to judge the market knowledge of the valuers involved. It can happen that an experienced valuer can "pull the wool over the eyes" of someone less experienced. The same can happen when one party is unrepresented. The position in respect of security of tenure is also important when assessing such agreements. A tenant at the expiry of a contracted out lease is in a potentially weak negotiating position and may be willing to pay a rent in excess of the normal open market level in order to avoid the cost of moving to new premises. As with all comparable evidence, it is essential for valuers to make enquiries and to be aware of factors behind the agreement.

3. Determination by an independent expert
   A determination by an independent expert is dependent upon the quality of the expert but the likelihood is that representations will have been received from both parties. Consideration of all the comparables put forward by the parties as well as any comparables known to the expert will have been undertaken. Such a determination ranks closely with an agreement between valuers in the weight that ought to be accorded to it.

4. An arbitrator's award
   The value of an arbitrator's award as evidence, generally, depends entirely upon the quality of the evidence put to him during the case, the manner in which it is expressed and the manner in which the parties were represented. Often an arbitrator may determine a figure in the light of the evidence that is not his personal opinion of the true market value. The weight to attach to the

award will therefore vary depending upon the quality of evidence, the arguments put before the arbitrator and the quality and market knowledge of the arbitrator and the surveyors acting on behalf of the parties.

5. Determination by the court under Part II of the Landlord and Tenant Act 1954
An arbitrator's award will rank ahead of a judge's determination of the rent payable under a new tenancy granted pursuant to Part II of the Landlord and Tenant Act 1954 because the arbitrator possesses the requisite expertise in the field with which to evaluate the parties' evidence and submissions.

6. Hearsay.

Example 32: Hotel Rabat

In this example, four hotels have recently been leased and have all been considered to some extent to be comparable to Hotel Rabat, the subject property.

Hotel Rabat is a four-star property located in the city centre with excellent visibility; it has 100 bedrooms, extensive conferencing facilities, good leisure facilities, good car parking and is well regarded in the marketplace. It is approximately 10 years old and trades mainly to corporate clientele, with a good selection of leisure guests and wedding business at weekends. The lease has an unexpired term of 15 years, with five yearly rent reviews on FRI terms to a shell and core specification.

The four comparables can be summarised as follows:

Hotel Canary is less well located than the subject property, close to the station just off the city centre in a slightly run-down part of the city. Surrounding the hotel are a number of cheaper hotels and hostels providing accommodation for more price conscious guests. The hotel caters to a mid-market clientele, with lower-rated corporate customers and individual leisure guests. It is about 80 years old, quite tired looking, with 60 bedrooms, a small bar and breakfast area and no car parking facilities. The hotel was let on a 15-year lease, shell and core basis, on FRI terms and the deal was completed two years ago. The rent equates to £7,000 per bedroom and is subject to review every five years.

Hotel Family is in a similar location to the Hotel Rabat and is less than one year old and provides top-end, four-star accommodation. It has 120 large bedrooms, excellent leisure facilities, two small meeting rooms, a fashionable restaurant (with a celebrity chef) and a top-rated spa. However, parking is limited. The hotel was let four months ago on a sale and leaseback on a fully fitted basis, paying 21% of turnover, of which £8,000 per bedroom was guaranteed. The guaranteed part of the rent was annually reviewed to RPI. The total rent is anticipated to be in the region of £12,500 per bedroom.

Hotel Cayman is located in the tourist section of the city and caters mainly to bus and coach tours. It has 400 bedrooms, a large restaurant and bar areas and is 40 years old. It was let 15 years ago on a 35-year lease on a fixed rent, with the last review a year ago showing a rent of £4,750 per bedroom, on FRI terms, on a shell and core specification, with reviews every seven years.

Hotel Dodecanese is a recently opened budget hotel with 100 bedrooms and is located on the outskirts of the city. It has minimal public areas, well-designed bedrooms and good parking. The property was let on a shell and core basis two months ago, on FRI terms, with annual RPI-linked rent reviews, and equates to a rent of £4,200 per bedroom. The hotel is part of a larger development that took two years to construct, and is now 80% open and the heads of terms for the lease were signed and agreed two years ago.

Methodology for Assessing Market Rent

Table 9.2 Summary of comparable hotels

Property	Date let	Rooms	Reviews	Rent/room (£)
Hotel Canary	2 years ago	60	5 yearly reviews	7,000
Hotel Family	4 months ago	120	Annual review, part to RPI, fully fitted	8,000–12,500
Hotel Cayman	1 year ago	400	Every 7 years, shell and core specification	4,500
Hotel Dodecanese	2 months ago	100	Annual review to RPI	4,200
Hotel Rabat (Subject)	Now	100	5 yearly rent reviews, shell and core specification	?

The above comparable evidence suggests that the rental value of the property will be somewhere between £4,200 and £12,500 per bedroom or between £420,000 and £1,250,000 per annum, depending on how the evidence is analysed.

It is essential that the valuer has sufficient experience and knowledge of the comparable transactions in order to accurately analyse the relevant comparable evidence.

In this example, the valuer was able to make the following adjustments:

Hotel Canary
- +2.5% for the size differential.
- +2.5% for the location.
- +2% for the age difference.
- +5% for the time difference between this letting and the valuation date.
- 0% for the difference in facilities at the hotel.
- 0% for the difference in the overall quality of the property.
- 0% difference for the nature of the rent review clause and the building specification.
- +12% total adjustment.

Hotel Family
- 0% for the size differential.
- 0% for the location.
- −10% for the age difference.
- 0% for the time difference between this letting and the valuation date.
- −10% for the difference in facilities at the hotel.
- −10% for the difference in the overall quality of the property.
- −10% difference for the nature of the rent review clause and the building specification.
- −40% total adjustment.

Hotel Cayman
- −7.5% for the size differential.
- +10% for the location.
- +2% for the age difference.
- +2.5% for the time difference between this letting and the valuation date.
- +2% for the difference in facilities at the hotel.
- +2% for the difference in the overall quality of the property.
- +2% difference for the nature of the rent review clause and the building specification.
- +13% total adjustment.

Hotel Dodecanese
- 0% for the size differential.
- +25% for the location.
- −10% for the age difference
- +10% for the time difference between the date the rent was agreed and the valuation date.
- +5% for the difference in facilities at the hotel.
- +10% for the difference in the overall quality of the property.
- −10% difference for the nature of the rent review clause and the building specification.

+35% total adjustment.

Table 9.3 Summary of adjustments for comparable hotels

	Rent/bedroom (£)	Adjustment (%)	Adjusted rent/bedroom (£)
Hotel Canary	7,000	+12	7,840
Hotel Family	12,500	−40	7,500
Hotel Cayman	4,750	+13	5,368
Hotel Dodecanese	4,200	+35	5,460

Source: Leisure Property Services

On this basis, the value of Hotel Rabat is likely to lie within £5,368–£7,840 per bedroom (£536,800–£784,000).

The valuer is then expected to use his judgment to determine where within that range this particular property would sit. In this instance, it has been determined that the value of Hotel Rabat would be £6,200 per bedroom, equating to £620,000.

It should be noted here that the above adjustments are specific to this example; all adjustments used in practice will be determined as much as possible by valuation evidence as well as by the personal experience of the individual valuer.

When are comparables used?

Comparables are used:

- Where trading is not the motivating factor behind the letting, for example in lifestyle properties.
- As a check for the profits method in most rental valuations.

It should be noted that almost all comparables will need to be adjusted to look at the underlying potential profitability of the hotel. As such, the key method for assessing hotel rents is the profit method of valuation.

Profits method

Almost without fail any operator looking to lease a hotel will review the trading potential of a particular unit before determining the level of rent that can be afforded. The usual methodology for assessing the rental value of a property is to determine the "divisible balance" and then apportion that amount between the landlord and the tenant.

Calculating the divisible balance involves working out the IBFC for the property and then deducting a number of items, including an annualised sum for the tenant's fixtures and fittings and a sum for working capital.

The apportionment of the divisible balance reflects the tenants bid. If the market is performing strongly and the property is desirable, the tenant will be prepared to offer more of the divisible balance to secure the property. Conversely, if the market is weak and there is less demand for the property the landlord will be prepared to accept a lower proportion of this profit for the property.

It is then normal practice to review this "rental bid" as a proportion of turnover, as a proportion of EBITDA and on a rent per bedroom basis in light of other market transactions.

Unfortunately, it is not always possible for the valuer to look through the trading accounts of the hotel when assessing the rental value of the property; ideally, the valuer would have access to such accounts but since few traditional leases provide for such information the assumed trading projections are normally down to the experience of the valuer.

As such it is be important for the valuer to ascertain the stabilised trading as described in Chapter 2. It may not always be possible for the valuer to interview the general manager of the hotel as sometimes the information provided may not be as full and frank as the valuer would hope for. Once again, experience in the local market is essential to help overcome such shortcomings.

Example 33: Hotel Jakarta

The valuer will need to take into account exactly what it is he is trying to value. The lease will detail the actual demise of the property (for example, does it include the FF&E) and will outline the terms on which the hotel will be valued.

The performance of the hotel is assessed by the valuer in today's values, based upon the demised premises; for example, if the hotel was let five years ago and was in good condition at the time of the letting, then the valuer needs to assess how that hotel would trade in today's market.

The process is very similar to that of a capital valuation all the way through to the IBFC line; the Occupancy Rate and ADR are assessed to calculate rooms' revenue, which is then worked into total revenue. Costs are then allocated across each department and in Unallocated Costs until the IBFC has been assessed. It is then standard to deduct fixed costs (except rent), including rates, management fees (where it is appropriate to do so), FF&E reserve (where it is appropriate to do so) and insurance costs.

Then the adjustments that are specific to rent review valuations are made; adjustments are made to the profit margins to take into account other items, for example, working capital and the cost of furnishing the property (assuming it has been let on a shell and core basis with no FF&E included).

The adjustment for fitting out should be made as a capital amount (probably as the cost per bedroom to provide a suitable level of FF&E for the hotel type) and then annualising it over a suitable time period (10 years seems to have come into favour as a conventional time period but there is no reason why this should not be altered if a more appropriate period can be shown).

It is deducted because the hotel will be unable to trade without FF&E and it is presumed that the hypothetical operator will need to invest the money to furnish the property. As such, the cost needs to be taken into account.

The working capital is also a cost of operation, and convention dictates that three weeks' revenue is deducted from the profit to make an appropriate allowance for this expense.

This works down to a divisible balance (which is lower than the EBITDA of the property because of the additional deductions). The divisible balance is the amount of profit available at the hotel that needs to be divided between the landlord and the tenant (between rent and retained profit).

Table 9.4 Rental value, Hotel Jakarta

Number of Bedrooms		60	
Occupancy Rate (%)		72.50	
ADR (£)			200.00
Total Rooms Revenue		£3,175,500	
Rooms Revenue/Total Revenue (%)		80.00	
Total Revenue		£3,969,375	
Direct Room Costs	32%	£1,016,160	
Other Revenue Costs	52%	£412,815	
Unallocated Costs	22.50%	£893,109	
IBFC		£1,647,291	41.50%
Fixed Costs			
Rates		£136,943	3.45%
Management Fee		£119,081	3.00%
Insurance		£43,663	1.10%
Total		£299,688	
Tenants FF&E			
Rooms		60	
@		£15,000	
Total		£900,000	£90,000
Working Capital	3 Weeks' Revenue	£229,002	
EBITDA			£1,188,828
Divisible Balance			£1,028,600
% as Rent	60.0%		£617,160
Rental Value			**£617,160**
Say			**£620,000**
Price per Bedroom			£10,333
% of Turnover			15.62
% of EBITDA			2.15
Rental Value			£620,000
Adjustment for Rent Review Period			−2.00%
Adjustment for Other Lease Terms			−10.00%
Say			**£545,600**

Source: Leisure Property Services

The decision on how to assess the proportion of divisible balance that the hypothetical tenant will be prepared to pay as rent to the landlord is quite difficult and will be based upon a number of things that will require the valuer to have a good knowledge of the local market. Prevailing economic conditions, demand for hotels from operators, hotel trading conditions (both the supply and demand inherent in the market) as well as the quality of the asset and its suitability and positioning within the local hotel market will all have an effect on the amount of the divisible balance that will be offered to the landlord. Needless to say, comparable evidence here is important, but needs to be analysed and used on a consistent basis.

Table 9.5 Rental valuation, Hotel London

Valuation Date	25/12/2005		
Number of Ensuite Bedrooms		42	
Occupancy Rate (%)		77.00	
ADR		115.00	
Number of Non-Ensuite Bedrooms		20	
Occupancy Rate (%)		77.00	
ADR		95.00	
Total Rooms Revenue (£)		1,891,467	
Rooms Revenue/Total Revenue (%)		85.00	
Total Revenue (£)		2,225,255	
Direct Room Costs (£)	27%	510,696	
Other Revenue Costs (£)	60%	200,273	
Unallocated Costs (£)	23.00%	511,809	
IBFC (£)		1,002,477	45.05%
Fixed Costs			
Rates		86,117	3.87%
Management Fee		66,758	3.00%
Insurance		22,253	1.00%
Total (£)		175,128	
Tenants FF&E	Rooms	42	
@		12,000	
Total		504,000	50,400
	Rooms	20	
@		6,500	
		130,000	13,000
Working Capital	3 weeks' revenue		128,380
EBITDA			738,340
Divisible Balance			635,570
% as Rent	55%		349,563
Rental Value (£)			**349,563**
Say (£)			**350,000**
Price per Bedroom (£)			5,645
% of Turnover			5.73
% of EBITDA			47.40

Source: Leisure Property Services

Further adjustments will then be made for specific lease terms that are unusual and for peculiarities in the rent review period.

It is then usual for the valuer to look at the result to assess whether it seems in line with the general market. There are three good tests for this; rent as a percentage of turnover, rent as a percentage of EBITDA and, to a lesser extent, rent per bedroom. This standback and review of the results is a vital part of the valuation process.

Sometimes it is less straight forward, as shown in the Hotel London example.

Valuation of Hotels for Investors

Table 9.6 Rental value — alternative approach, Hotel London

Valuation Date	25/12/2005		
Number of Ensuite Bedrooms		62	
Occupancy Rate (%)		77.00	
ADR		115.00	
Number of Non-Ensuite Bedrooms		0	
Occupancy Rate (%)		77.00	
ADR		95.00	
Total Rooms Revenue		2,003,887	
Rooms Revenue/Total Revenue (%)		85.00	
Total Revenue (£)		2,357,514	
Direct Room Costs (£)	27%	541,049	
Other Revenue Costs (£)	60%	212,176	
Unallocated Costs (£)	23.00%	542,228	
IBFC (£)		1,062,060	45.05%
Fixed Costs (£)			
Rates (£)		91,236	3.87%
Management Fee (£)		70,725	3.00%
Insurance (£)		23,575	1.00%
Total (£)		185,536	
Tenants FF&E			
	Rooms	62	
@		12,000	
Total (£)		744,000	74,400
Other Improvements		1	
@		350,000	
		350,000	35,000
Working Capital (£)	3 weeks' revenue		136,010
EBITDA			782,223
Divisible Balance (£)			631,113
% as rent	55%		347,112
Rental Value (£)			**347,112**
Say (£)			**350,000**
Price per Bedroom (£)			5,645
% of Turnover			14.85
% of EBITDA			44.74

Source: Leisure Property Services

Example 34: Hotel London

In this example, the hotel was let 35 years ago on an 84-year lease and since that time substantial tenants' improvements, including air-conditioning and ensuite facilities, have been undertaken to make the bedrooms more attractive to guests. However, the rent review clause specifically disregards any tenants' improvements from being included when assessing the market rent.

At the start of the lease, the hotel had 42 non-ensuite rooms and 20 ensuite rooms; the present bedroom configuration is 48 ensuite bedrooms.

In this instance, there are two approaches the valuer will have to adopt to accurately assess the rental value of the property. The first approach will be to carry out a calculation on the property, as originally it had 62 rooms.

The second approach will be to assume the improvement works have been carried out at the expense of the "hypothetical tenant" and to capitalise this expenditure.

In this instance, both methods provide the same value. In the event of significant differences the valuer will need to determine which method would most accurately represent the prevalent market approach.

When is the profits method used?

The profits method of valuation is used at almost every rent review (if the review is to "market rent") and new letting where the reason behind a potential tenant's occupation is to operate a financially successful business. This valuation method is the most effective way of determining the rental level of a property. However, if the lease determines a different method of valuation (for example, it is assessed to 4% of the capital value or assessed to 80% of office rental on the adjacent building), then the specified methodology will need to be adhered to.

However, it is not always used when assessing lifestyle properties, as the motivation behind occupation has nothing to do with running a successful business and may therefore attract bids that are calculated without reference to profitability.

There are two other methods of assessing rents that are sometimes used and these are summarised below.

Rental levels as a percentage of total investment

In certain circumstances, the rental level of a hotel may be initially based on building costs where an operator will offer a certain guaranteed return based on total investment costs. These initial offers are always reviewed to ensure the rent is "affordable" by the tenant.

Example 35: Hotel Ljubljana

A developer has secured a site for a new hotel and is building a 210-bedroom hotel, which on completion, will be let to a well-respected operating company. The land costs £2,000,000, the construction of the hotel to shell and core specification will cost an additional £19,000,000 and, with fees, finance expenses etc, the overall cost of the development is expected to be £23,800,000. The operator has offered to provide a rent based on a 6.75% gross return of the total investment cost, to be annually reviewed to RPI (or £1,606,500 per annum).

Table 9.7 Rental calculation, Hotel Ljublijana (£)

Cost of Land	2,000,000
Cost of Building	19,000,000
Other Costs	1,800,000
Total Costs	23,800,000
Gross Return Required (%)	6.75
Rent	1,606,500

Source: Leisure Property Services

$$\text{Rent} = 6.75\% \times £23{,}800{,}000 = £1{,}606{,}500$$

It should be noted that the agreed return will generally be higher than the market price for an operational investment with the same operator. This reflects the added development risk.

In development scenarios such as this, the developer will normally try and ascertain rental bids from similar quality operators (in terms of covenant strength) based on a percentage of the total cost of developing the hotel.

This method of assessing rents is mainly used when considering the level of rent the operator should offer prior to the development being completed. It is also used as a check method for market rents for other properties.

Rents per metre or square foot

In some instances, the rent of a hotel is also analysed on a price per metre or square foot basis although, at the current time this is not widespread practice and is not commonly used to agree rental levels.

However, there has recently been an increase in traditional property investors moving into the hotel market, and with them has come more analysis of hotel returns on a floorspace basis. It is therefore quite possible that comparable evidence for hotels could be analysed on a per square foot basis in the future.

Summary

When assessing the market rental value of a hotel the intention is to reflect what would happen if the property were placed on the market at that time. Although the prevalent method of assessment is the profits method of valuation, it is essential that the valuer refers as much as possible to market evidence. Unfortunately however, most comparable evidence that is provided as rental values per bedroom is too superficial in terms of detail to be more use than a simple benchmark for the tone of the market.

Methodology for Calculating Rateable Values in the UK

Introduction

Most countries apply a level of taxation on commercial property and the methods of assessing this level of tax varies throughout the world. This chapter specifically relates to the UK hotel market, where commercial property taxation is known as rates.

What are rates?

In the UK, all non-domestic properties have a rateable value that is used to assess the level of property tax that is paid to the local government (in contrast to income tax, corporation tax and capital gains tax which go towards to central government).

The rateable value (RV) is based on an assessment of the annual rent of a similar property, assuming it was available to let on the open market at a fixed valuation date. The rating lists that came into effect on 1 April 2005 (the most recent revaluation) are based on a valuation date of 1 April 2003. All properties are revalued every five years.

The RV of the hotel is not the amount that is paid in business rates, but it is the basis for the calculation of your business rates bill. The local authority calculates the bill by multiplying the RV by a factor set by central government each year. This is known as the multiplier of uniform business rate (UBR).

Assessing the taxation liability of a hotel
Example 36: Hotel Sofia

Hotel Sofia has been given a rateable value of £750,000 and the UBR for the relevant year was 43.3 pence in the pound, so the rates payable was £324,750

$$£750,000 \times .433 = £324,750$$

Transitional relief

In some instances, a new rating assessment could lead to a substantial increase in the taxation that is required, which could cause undue hardship to the hotelier.

Following the rating revaluation, which took effect on 1 April 2005, a new system of transitional relief was implemented. This restricted increases and decreases in rate liability over the first three years of the new assessment. The maximum amount by which the rate liability of a property may increase in 2006/07 was 17.5% plus an adjustment for inflation of 2.6%.

Some existing hotels therefore benefit from transitional relief and this needs to be taken into account when determining the annual cost of the rating liability for the property. Transitional relief softens the impact of revaluation by phasing in the changes to the rates bill over a period of time. Different transitional arrangements apply, depending on whether your bill has increased or decreased and whether your business is classed as small or large.

A small business is defined as a property with a rateable value of less than £15,000 outside London and less than £21,500 within London.

Table 10.1 shows the proposed limits by which a rates bill can increase in a single year before transitional arrangements apply:

Table 10.1 Rates assessments: transitional increases

	Small business (%)	Large business (%)
2005/06	5	12.5
2006/07	7.5	17.5
2007/08	10	20
2008/09	15	25
2009/10	n/a	n/a

Source: Leisure Property Services

Slightly different arrangements apply for the proposed limits by which a rates bill can decrease in a single year, as shown in Table 10.2.

Table 10.2 Rates assessments: transitional decreases

	Small business (%)	Large business (%)
2005/06	30	12.5
2006/07	30	12.5
2007/08	35	14
2008/09	60	25
2009/10	n/a	n/a

Source: Leisure Property Services

Example 37: Hotel Stockholm

Hotel Stockholm had an RV of £350,000 after the 2000 rating assessments and in 2005 this increased to £620,000.
In 2005, the rates liability for the hotel was £147,700, calculated off the RV of £350,000 with a UBR of £42.2.
In 2006, the strict liability would have been £268,460, based on an RV of £620,000 and a UBR of 43.3p. However, this would involve an increase in the rates liability of £120,760 (or 81.7%). As such, the phasing provisions would come into effect and the rates liability for 2006/07 would be reduced as follows:
2005/06 rating liability £147,700. Increased by 17.5% phasing provision (£173,547.50) and then increased by inflation of 2.6%, providing an end rating liability of £178,059.73 in 2006/07.

Method of assessing rateable value

Rateable values are assessed on behalf of the government by the valuation office (VO) and it is based on the notional rental value of the property. However, the difficulties in assessing rental values due to the lack of direct comparables, and the reliance upon the profits method of valuation in assessing hotel rental values, has resulted in a relatively straight forward valuation methodology being agreed by the VO based on revenue generated by the hotel for the 2005 rating list.

In order to try and determine the profitability at the unit the VO has come up with a number of factors that adjust the fair maintainable receipts (FMR) per bedroom. Other revenue-generating areas are calculated as equivalent double bedded units (EDBUs) total to arrive at an adjusted double bedded unit (ADBU) total.

The FMRs for the 2005 list was based upon the perception of trading in 2003 and, as such, was set at this time. Each hotel had a duty to provide the VO with turnover details at this time and so 2003 turnover numbers should always provide the starting point for the valuer when calculating the FMR.

The valuer needs to determine the location, age and quality (the equivalent AA rating assessment) of the property as different scales apply to each of these factors. The VO has provided guidance on the relevant scales, as detailed in Table 10.3.

Table 10.3 Four-star hotels (and equivalent)

Accommodation Receipts Per DBU (£)	Guide % to RV					
	Post-1984		1960–84		Older hotels	
	Minimum	Maximum	Minimum	Maximum	Minimum	Maximum
27,500	11.5	13	10	11.5	9.5	11
25,000	11.125	12.75	9.625	11.125	9	10.375
22,500	10.5	12.25	9	10.5	8.25	9.5
20,000	9.875	11.5	8.5	9.875	7.75	9
17,500	9	10.5	8	9.25	7.375	8.5
15,000	8.125	9.625	7.5	8.75	7.125	8.125
12,500	7.5	9	7	8.25	6.75	7.75
10,000	7	8.5	6.5	7.875	6.25	7.25
7,500	6.75	8.25	6.25	7.75	6	7

Source: Leisure Property Services

Valuation of Hotels for Investors

Table 10.4 Three-star hotels (and equivalent)

Accommodation Receipts Per DBU (£)	Guide % to RV					
	Post-1984		1960–84		Older hotels	
	Minimum	Maximum	Minimum	Maximum	Minimum	Maximum
27,500	11.75	13.25	10.75	12.25	10.25	11.75
25,000	11.5	13	10.25	11.75	9.75	11.25
22,500	10.875	12.5	9.375	10.875	8.875	10.25
20,000	10.125	11.75	8.75	10.125	8.25	9.5
17,500	9.25	10.75	8.25	9.5	7.75	9
15,000	8.5	9.875	7.75	9	7.25	8.5
12,500	8	9.25	7.25	8.5	6.75	8
10,000	7.375	8.75	6.5	8	6.25	7.5
7,500	6.5	8	5.5	7.25	5.5	6.75
5,000	6	7.5	5	6.75	5	6.25

Source: Leisure Property Services

Table 10.5 Two-star hotels (and equivalent)

Accommodation Receipts Per DBU (£)	Guide % to RV					
	Post-1984		1960–84		Older hotels	
	Minimum	Maximum	Minimum	Maximum	Minimum	Maximum
20,000	11.75	13.25	9	11	8.5	10.5
17,500	11.5	13	8.75	10.75	8.25	10.25
15,000	11	12.5	8.25	10.25	7.75	9.75
12,500	10.5	12	7.75	9.75	7.25	9.25
10,000	9.75	11.25	7.25	9.25	6.75	8.75
7,500	8.75	10.25	6.625	8.625	6.25	8.25
5,000	7.875	9.25	5.875	7.875	5.75	7.75
2,500	7.5	8.75	5.5	7.5	5.5	7.5

Source: Leisure Property Services

The valuer must also determine the amount of EDBUs at the hotel. Once again, these are scales that apply as follows:

Table 10.6 EDBU factors

Room type	En-suite	Size guide
Double/twin	1	10–20m^2
Single	0.7	Up to 10m^2
Family	1.25	20–30m^2
Suite — standard	1.5	2 rooms or over 30m^2
Suite — superior	2	2 rooms or over 30m^2

Source: Leisure Property Services

The FMR will then be divided by the ADBUs to determine which percentage band the hotel falls into, with adjustments being available within that bracket to reflect the potential profitability of the hotel. The final step is to review the rateable value to check that this methodology has produced an appropriate result.

Example 38: Hotel Bratislava

Hotel Bratislava is a three-star, mid-market hotel built in 1983. It has 100 bedrooms, which are all arranged on the ground and first floor, along with a small bar and restaurant of 200m^2. There are no other facilities, and the hotel generated £1,100,000 in turnover in 2003.

Table 10.7 Calculation of EBUs, Hotel Bratislava

Accommodation		EBUs
100 bedrooms @ 20m^2	(×1)	100
Bar/restaurant 200m^2	(20m^2 = 1EBU)	10
Total		110

Source: Leisure Property Services

Total FMR = £1.1m equates to £10,000 per EBU
(£1,100,000/110 = £10,000)

The agreed scale is between 6.5% (£1,100,000 × 6.5% = £71,500) and 8% (£1,100,000 × 8% = £88,000).
 Bearing in mind all the factors relating to the hotel, it was agreed that the appropriate scale was 7%, equating to an RV of £77,000.

Example 39: Hotel Amman

In this example, we have a 100-bedroom, mid-market hotel, which was built in 1995 and earning £3.3m revenue per year. It has 1,000m^2 of meeting space, 500m^2 of bar and restaurant facilities and 1000m^2 of leisure space. The property is arranged on lower-ground, ground and two upper floors, with 10 guest bedrooms arranged on the lower-ground floor. The hotel has been graded as four star by the AA.

Table 10.8 Calculation of EBUs, Hotel Amman

Accommodation		EBUs
1,000m² meeting space	(20m² = I EBU)	50
500m² bar and restaurant	(20m² = 1EBU)	25
100mm² leisure club		15
4 standard suites @ 34m²	(× 1.5)	6
2 junior suites @ 28m²	(× 1.25)	2.5
84 standard rooms @ 20m² on ground, first and second floors	(× 1)	84
10 standard rooms @ 20m² on the lower ground floor	(× 0.85)	8.5
Total		191

Source: Leisure Property Services

The total FMR is calculated at £3.3m, the same as the actual total revenue. This equates to £17,277 per EBU (£3,300,000/191).

The agreed scale for a property of that type and age is between 8.125% and 9.625%. The size of the property, the age and condition of the property and the layout of the property were all discussed and it was agreed that the appropriate percentage for this property was 9.1%.

This results in a rateable value for the property of £300,300 (£3,300,000 × 9.1%).

Summary

The UK property regime is based upon revaluations every five years. Almost every revaluation has been based on slightly different guidelines and parameters and it is essential that the valuer considers the most up-to-date guidelines when reviewing the rateable assessment of a property. Most countries have different ways of assessing property taxation and the valuer must look into local practices and take local advice when assessing such property taxes.

Site Values and How They are Determined

Introduction

This chapter will explain how the value of a hotel site is determined, and how hoteliers and developers decide where new hotels should be built and indeed whether the development is feasible.

The theory is simple; new hotels will be considered feasible by developers if they are able to be sold for more money than they cost to build. Hoteliers will normally have different criteria but would not tend to develop the property if the end value is less than the development costs.

Basic methodology

The basic methodology behind a site valuation is twofold. The valuer will look at any comparable transactions to refer to a price for the land first and foremost to work out the range in which it is likely that the value will sit.

Unfortunately, because of the specialist nature of a hotel's trading profile and as there are rarely sufficient comparable site transactions to refer to, it is usually necessary for a valuer to resort to undertaking a residual valuation.

During the consideration of several valuation cases, the Lands Tribunal (the court that decides disputes in UK property matters) have raised various concerns over the adequacy and accuracy of the residual method of valuation. This is due to the large number of variables used in arriving at the end value, and the sensitivity of the end value to even minor changes in these variables.

As such, it is important that the valuer highlight in his report these potential inadequacies and draw the client's attention to the uncertainty in the resulting end value.

That said those active in the market do tend to carry out residual valuations to determine the price that can be paid for a site and, as such, it is probably the best method for arriving at the value of the site as long as the valuer is aware of the variables used by the majority of the active purchasers in the market at the time of the valuation.

In simple terms, the residual method of valuation can be summarised as follows:

Valuation of Hotels for Investors

Likely future value of the completed development
Costs of development
Developer's profit
Gross land value

The detail behind this methodology is illustrated by the Hotel Harare example below:

Example 40: Hotel Harare

The likely future value (LFV) for the hotel is calculated at £20,000,000. The cost of the development (including finance) is £14,000,000. The developer's profit is assessed to be 25% of development costs, which is £3,500,000. Therefore, the gross land value is £2,500,000.

$$£20{,}000{,}000 - £14{,}000{,}000 - £3{,}500{,}000 = £2{,}500{,}000.$$

The LFV will be the estimated sale price of the completed development and will be calculated taking into account the projected trading profile of the hotel throughout the buildup to stabilised trading and then multiplied by the appropriate multiple.

The development costs will need to include the following:

- Demolition works.
- Site remediation works.
- Construction of the hotel.
- Ground works.
- Fees for professional advisors, including architects, quantity surveyors, structural engineers, project managers etc.
- Finance costs.
- Transaction costs.
- FF&E.
- Pre-opening expenses.

The developer's profit will be the return that the developer will require to carry out the development and is usually expressed as a percentage of the total development costs (including the land), although it is sometimes referred to as a percentage of the LFV.

The gross land value then needs to incorporate the purchase and finance costs to arrive at the net land value. This value is then usually calculated on a per bedroom basis to see if it is in line with usual market values.

In *Singer & Friedlander Ltd* v *John D Wood & Co* [1977], a negligence case was heard regarding a site valuation. The presiding judge said:

> The factors and activities which a competent Valuer will consider and undertake in valuing land cannot be composed in such a way as to indicate an unvarying approach to every problem which confronts him. But a collection of them from which he will choose at will in a given circumstance must include the following:
>
> - The kind of development of the land to be undertaken.
>
> - The existence, if any, of planning permission in outline or in detail as to a part or the whole of the land. And if permission be for the building of houses, the situation and acreage of part of the land excluded

from planning permission because, for example, of a tree preservation order, the need for schools and the lay-out of roads and other things. Furthermore, the number of houses permitted or likely to be permitted to be built by the planning authority is a relevant, indeed a vital factor.

- The history of the land, including its use, changes in ownership and the most recent buying prices, planning applications and permissions, the implementation or otherwise of existing planning permissions, the reason for the failure, if it be the fact that a planning permission has not been implemented.

- The position of the land in relation to surrounding countryside, villages and towns and places of employment; the quality of access to it, the attractiveness or otherwise of its situation.

- The situation obtaining about the provision of services, for example, gas, electricity, sewage and other drainage and water.

- The presence, if it be so, of any unusual difficulties confronting development which will tend to increase the cost of it to an extent which affects the value of the land. A visit to the site must surely find a prominent place among physical activities to be undertaken.

- The demand in the immediate localities for houses of the kind likely to be built, with special regard to the situation of places of employment and increases to be expected in demand for labour. This will involve, inevitably, acquiring knowledge of other building developments recently finished or still in progress, especially having regard to rate of disposal, density and sale price of the houses disposed of. In this way the existence, if any, of local comparables, a valuable factor, can be discovered.

Consultation with senior officers of the local planning authority is almost always regarded as an indispensable aid, likewise a knowledge of the approval planning policy for the local area; a study of the approved town or county map may prove rewarding.

Whether ascertaining from the client if there have been other previous valuations of the land, and to what effect, should be undertaken is probably questionable because a Valuer's mind should not be exposed to the possibilities of affectation by the opinions of others.

If he is a man whose usual professional activities do not bring him regularly into the locality, or what is more important has never done so, he will obviously need to be especially careful in collecting as much relevant local knowledge as he can, possibly by consulting Valuers who work regularly in the area.

The availability of a labour force which can carry out the prospective development.

When this harvest of knowledge has been gathered in by the Valuer with, on occasions, as must be accepted, help from a competent member or members of his staff or the firm for whom he works, he must assess the worth of the details of it. Some of it he may set aside either as being unreliable or for some other good reason. Other details may impress him as reliable as facts which speak for themselves or upon which he can, using his best judgment, rest assumptions without the making of which his task is rendered impossible. With these he has to try to penetrate the mists of the future and withal bring to bear upon him his training, skill and experience in order to produce a carefully achieved conclusion in terms of monetary value.

What are the final steps to be taken before pronouncing a figure? Having decided upon his basic facts and made his assumptions, he collects together from his own resources, professional publications, local comparables and elsewhere the current market trends in the selling prices of the houses, the cost of borrowing money in the

Table 11.1 Forecast in present values, Hotel Bucharest

Year	1		2		3	
No of Rooms	220		220		220	
Rooms Sold	48,180		56,210		60,225	
Rooms Available	80,300		80,300		80,300	
Occupancy (%)	60.0		70.0		75.0	
ADR	80.00		90.00		100.00	
RevPAR	48.00		63.00		75.00	
Growth (RevPAR) (%)			31.3		19.0	
Revenues ('000s)		(%)		(%)		(%)
Rooms	3,854.4	70.0	5,058.9	67.5	6,022.5	65.0
Food	660.8	12.0	1,049.3	14.0	1,389.8	15.0
Beverage	550.6	10.0	786.9	10.5	1,111.8	12.0
Total Food & Beverage	1,211.4	22.0	1,836.2	24.5	2,501.7	27.0
Room Hire	0.0	0.0	0.0	0.0	0.0	0.0
Leisure Club	330.4	6.0	449.7	6.0	555.9	6.0
Other Income	110.1	2.0	149.9	2.0	185.3	2.0
Total Revenue	**5,506.3**	**100.0**	**7,494.7**	**100.0**	**9,265.4**	**100.0**
Departmental Profit						
Rooms	2,505.4	65.0	3,642.4	72.0	4,456.7	74.0
Total Food & Beverage	363.4	30.0	642.7	35.0	1,000.7	40.0
Room Hire	0.0	100.0	0.0	100.0	0.0	100.0
Leisure Club	79.3	24.0	107.9	24.0	133.4	24.0
Other Income	35.2	32.0	48.0	32.0	59.3	32.0
Total Departmental Profit	**2,983.3**	**54.2**	**4,441.0**	**59.3**	**5,650.0**	**61.0**
Departmental Costs	**2,523.0**	**45.8**	**3,053.7**	**40.7**	**3,615.4**	**39.0**
Undistributed Operating Expenses						
Administrative & General	550.6	10.0	674.5	9.0	787.6	8.5
Sales & Marketing	330.4	6.0	374.7	5.0	370.6	4.0
Property Operations & Maintenance	220.3	4.0	299.8	4.0	370.6	4.0
Utility Costs	176.2	3.2	239.8	3.2	296.5	3.2
Total Undistributed Expenses	**1,277.5**	**23.2**	**1,588.9**	**21.2**	**1,825.3**	**19.7**
Income Before Fixed Costs	**1,705.8**	**31.0**	**2,852.1**	**38.1**	**3,824.8**	**41.3**
Fixed Costs						
Reserve for Renewals	110.1	2.0	224.8	3.0	370.6	4.0
Property Taxes	258.8	4.7	352.2	4.7	435.5	4.7
Insurance	55.1	1.0	74.9	1.0	92.7	1.0
Management Fees	165.2	3.0	224.8	3.0	278.0	3.0
Total Fixed Costs	**589.2**	**10.7**	**876.9**	**11.7**	**1,176.7**	**12.7**
EBITDA	**1,116.7**	**20.3**	**1,975.2**	**26.4**	**2,648.0**	**28.6**

Source: Leisure Property Services

market, the average value of land per acre on a national basis and on a local basis, and the average value of land per plot on like basis covering a period of a few months preceding the valuation. So the exercise in valuation proceeds in a fairly conventional way, which I call the "discount method," and which is not the only way, of course, as follows:

(a) Assume, if it is not actually known, the number of houses to be built on site—the density.
(b) Assume the rate of building which bears relation to the anticipated rate of sales.
(c) Determine the number of years for the completion of the development.
(d) Adopt a deferment rate.
(e) Alight upon a plot value and/or a value per acre. How? The answer seems to be by using local comparables or median plots values for the area or both; perhaps neither of these aids will be present.

It should be noted that this methodology is being used to assess the market value. As such, it is important that the valuer looks at the costs and the end value in terms of how the market would assess these items.

Calculating the likely future value

It is more difficult for a valuer to determine the trading profile for a new hotel compared with an existing hotel at some point in the future, as there is no trading history to base future earnings on, and the location, style, design, corporate environment and staffing levels are all relative unknowns.

When looking to develop a new hotel it is usual for a developer to commission a feasibility study to ensure that the development makes sense, and to ascertain, with some degree of certainty, where the trading of the hotel is likely to come out. It is beyond the brief of this book to explain in any detail the methodology behind such feasibility studies but, in essence, it is similar to the process outlined in Chapter 2 although with less factual information to start with.

Most developers commission feasibility studies and therefore it is perfectly sensible for the valuer to base the valuation upon such a study, as that is the normal market practice.

The valuer will review the feasibility study, casting an expert eye over the commentary and projections to see if they tally with his experience of the local hotel market. In some instances, the valuer will have to make adjustments to the projections to tie in with what he feels the market would expect in such a location and from such a hotel.

Example 41: Hotel Bucharest

In this example, Hotel Bucharest is anticipated to trade as shown in Table 11.1.

Table 11.2 Income Capitalisation, Hotel Bucharest

Net Base Cash Flow			
EBITDA in 1 values in year 3			2,648,047
Capitalised at 9.25%		10.81	28,627,534
Less	Income Shortfall	2,204,200	
	Gross Value		26,423,335
Say			26,400,000

Source: Leisure Property Services

Valuation of Hotels for Investors

Equal consideration is then required of the development costs, as shown in Table 11.3.

Table 11.3 Development costs, Hotel Bucharest

	£ Total	(£)
Date of Valuation 01/04/2006		
Value at Completion		26,400,000
Construction Costs		
Structure, Shell & Core	4,925,123	
Internal Secondary Works	1,788,555	
M&E Works	4,021,000	
Fitting Out Works	345,000	
Finishes	680,000	
External Works	100,000	
Preliminaries	2,555,125	
Connection to Utilities	146,250	
Contingency	3,124,025	
Subtotal	17,685,078	
Construction Fees & Insurance		
Architect	350,000	
Employer's Agent	70,000	
Project Management	115,400	
Insurance & Local Authority Fees	105,250	
Contingency	42,150	
Subtotal	682,800	
FF&E		
The associated costs to be met by the tenant		
Pre-opening Costs	200,000	
Subtotal	18,567,878	
Other Expenses & Fees		
Operator's Technical Assistance Fees	152,500	
Legal Fees	62,500	
Financial Costs	1,250,000	
Subtotal	1,465,000	
Land Costs		
Rights of Access Insurance	100,000	
Legal, Surveying & Planning Consultants' Costs	300,000	
Subtotal	400,000	
Total Development Cost	20,432,878	£92,877 per bedroom
Finance Costs	6.5%	664,069
Developers' Profit	15.0%	3,064,932
Total Costs		**24,161,878**
Gross Site Value		2,238,122
Net of Stamp Duty	4.0%	2,152,040
Site Finance @ 6.5% for 1.5 year		218,217
Stamp Duty		86,082
Site Value		1,933,823
Say		1,935,000
Per Bedroom		9,258

Source: Leisure Property Services

Site Values and How They are Determined

Table 11.4 Analysis of building costs, Hotel Bucharest

Bedrooms					220
Size					12,120m²
	£ Total	£/key	£/m²	% of Total Construction Costs	% of Total Overall Cost
Construction Costs					
Structure, Shell & Core	4,925,123	22,387	406	27.85	20.38
Internal Secondary Works	1,788,555	8,130	148	10.11	7.40
M&E Works	4,021,000	18,277	332	22.74	16.64
Fitting Out Works	345,000	1,568	28	1.95	1.43
Finishes	680,000	3,091	56	3.85	2.81
External Works	100,000	455	8	0.57	0.41
Preliminaries	2,555,125	11,614	211	14.45	10.58
Connection to Utilities	146,250	665	12	0.83	0.61
Contingency	3,124,025	14,200	258	17.66	12.93
Subtotal	17,685,078	80,387	1,459	100.00	73.19
Construction Fees & Insurance					
Architect	350,000			62.30	2.28
Employer's Agent	70,000	318	6	7.93	0.29
Project Management	115,400	525	10	13.07	0.48
Insurance & Local Authority Fees	105,250	478	9	11.92	0.44
Contingency	42,150	192	3	4.77	0.17
Subtotal	682,800			100.00	3.65
FF&E	The associated costs to be met by the tenant				
Pre-opening Costs	200,000				
Subtotal	18,567,878	84,399		1,532	76.85
Other Expenses & Fees					
Operator's Technical Assistance Fees	152,500	693	13	10.41	0.63
Legal Fees	62,500	284	5	4.27	0.26
Financial Costs	1,250,000	5,682	103	85.32	5.17
Subtotal	1,465,000	6,659	121	100.00	6.06
Land Costs					
Rights of Access Insurance	100,000	455	8	25.00	0.41
Legal, Surveying & Planning Consultants' Costs	300,000	1,364	25	75.00	1.24
Subtotal	400,000	1,818	33	100.00	1.66
	–	–			0.00
Total Development Cost	20,432,878	92,877		1,686	84.57
Finance Costs	6.5%	664,069		17.81	2.75
Developers' Profit	15%	3,064,932		82.19	12.68
Total Costs		24,161,878			100.00

Source: Leisure Property Services

It is important that these building costs are analysed to ensure that they are appropriate for the sort of development that is proposed, and indeed that the building costs are typical of the sort of costs that would be expected within the market.

Any savings that a particular developer can bring to a development that are not typical of the market need to be discounted when carrying out the residual valuation (if the purpose of the exercise is to discover the market value).

Analysis of the building costs is normally carried out on a per key or per square meter basis, as shown in Table 11.4. In this example, contingency allowances have been specifically allocated to the construction costs and on construction fees, which totals close to 13% of the total development costs.

It is agreed that the Finance Costs would be 6.5% per annum, which is the prevalent borrowing rate for developments of this type. In this instance, it has been assumed that because of the phasing of the development programme, the borrowing requirement will equate to half of the money being needed at the start of the development process, with the remainder being required at a later stage.

Developer's profit

This is one of the more contentious items, as potentially each developer has his own criteria for carrying out developments. Certain hoteliers are less concerned with development profits than with getting new hotels built, which effectively reduces the amount of development profit that the valuer needs to incorporate into the valuation. However, it will be down to the experience of the valuer to determine what level of developer's profit to deduct from the likely future value.

Land value

To determine the value of the site, the valuer must deduct the cost of financing the purchase from the day when such costs will accrue and include the transaction costs to arrive at the net site value.

It should be reiterated that this method is susceptible to minor changes in the variables and, as such, the valuer should place as much emphasis on comparable transactions as possible.

In the example of Hotel Bucharest, the site value equated to £1,935,000, which breaks back to £9,258/room. It is at this point that the valuer needs to stand back and see if the end result is sensible.

Reference needs to be made to any other site transactions that are relevant. It is sometimes useful to review the land value as a proportion of the overall end costs and value. In this instance, £9,250/room equates to approximately 10% of the development cost and 7.3% of the end value, which is deemed normal in this particular area. The proportion of the overall value that the land value equates to will normally relate directly to the scarcity of supply of suitable sites, with hotel developments in Hong Kong showing a higher element of land value as a percentage of the completed development than for a resort hotel in Cape Verde, another newly emerging location.

Summary

Assessing site values is usually carried out through the use of a residual valuation because most potential developer's look to buy land on this basis. However, the array of assumptions required and the susceptibility of the end value to even minor changes in these assumptions means that it is essential that warnings are delivered to the client as to the potential inaccuracy and unreliability of this valuation method.

Valuation with Special Assumptions for Secured Lending Purposes

Introduction

As part of the *Red Book*'s requirements for secured lending purposes, the valuer is required to carry out a valuation subject to special assumptions, or what was previously referred to as a forced sale valuation.

The assumptions can vary depending upon the requirements of the client, but usually they will include:

- A shortened time frame to complete the marketing and the disposal of the property.
- Some concerns as to the trading potential of the unit, for example, the loss of the operating licences.
- Some concerns as to the accuracy of the trading information that has been provided, for example that the past accounts have not been provided, or cannot be relied upon.

In Chapter 5, we outline the *Red Book*'s suggested special assumptions.

The nature of the assumptions will dictate the methodology that the valuer should use to determine the market value subject to special assumptions, which can be used as a "pick list", although any special assumptions can be requested.

The most accurate method to assess this 'notional' value is to build up the trade to a stabilised level of performance over the relevant number of years, and then to multiply the EBITDA by the relevant multiple, less any relevant capital expenditure. The special assumptions that are specified will determine the starting point for the trading profile of the hotel.

Example 42: Hotel Nairobi

In this case, the special assumptions are as follows:

- Restricted marketing period of three months.
- No accounts available.
- The hotel is closed and is not trading.
- The hotel has lost the justices' licence.

The market value of the hotel without the special assumptions has been assessed in Table 12.1.

Table 12.1 Forecast in present values, Hotel Nairobi

Year	1		2		3	
No of Rooms	100		100		100	
Rooms Sold	27,375		27,375		27,375	
Rooms Available	36,500		36,500		36,500	
Occupancy (%)	75.0		75.0		75.0	
ADR	80.00		80.00		80.00	
RevPAR	60.00		60.00		60.00	
Growth (RevPAR) (%)			0.0		0.0	
Revenues ('000s)		(%)		(%)		(%)
Rooms	2,190.0	65.0	2,190.0	65.0	2,190.0	65.0
Food	505.4	15.0	505.4	15.0	505.4	15.0
Beverage	404.3	12.0	404.3	12.0	404.3	12.0
Total Food & Beverage	909.7	27.0	909.7	27.0	909.7	27.0
Room Hire	0.0	0.0	0.0	0.0	0.0	0.0
Leisure Club	202.2	6.0	202.2	6.0	202.2	6.0
Other Income	67.4	2.0	67.4	2.0	67.4	2.0
Total Revenue	**3,369.2**	**100.0**	**3,369.2**	**100.0**	**3,369.2**	**100.0**
Departmental Profit						
Rooms	1,620.6	74.0	1,620.6	74.0	1,620.6	74.0
Total Food & Beverage	363.9	40.0	363.9	40.0	363.9	40.0
Room Hire	0.0	100.0	0.0	100.0	0.0	100.0
Leisure Club	84.9	42.0	84.9	42.0	84.9	42.0
Other Income	23.6	35.0	23.6	35.0	23.6	35.0
Total Departmental Profit	2,093.0	62.1	2,093.0	62.1	2,093.0	62.1
Departmental Costs	1,276.3	37.9	1,276.3	37.9	1,276.3	37.9
Undistributed Operating Expenses						
Administrative & General	286.4	8.5	286.4	8.5	286.4	8.5
Sales & Marketing	134.8	4.0	134.8	4.0	134.8	4.0
Property Operations & Maintenance	134.8	4.0	134.8	4.0	134.8	4.0
Utility Costs	107.8	3.2	107.8	3.2	107.8	3.2
Total Undistributed Expenses	663.7	19.7	663.7	19.7	663.7	19.7
Income Before Fixed Costs	**1,429.2**	**42.4**	**1,429.2**	**42.4**	**1,429.2**	**42.4**
Fixed Costs						
Reserve for Renewals	134.8	4.0	134.8	4.0	134.8	4.0
Property Taxes	158.4	4.7	158.4	4.7	158.4	4.7
Insurance	33.7	1.0	33.7	1.0	33.7	1.0
Management Fees	101.1	3.0	101.1	3.0	101.1	3.0
Total Fixed Costs	427.9	12.7	427.9	12.7	427.9	12.7
EBITDA	**1,001.3**	**29.7**	**1,001.3**	**29.7**	**1,001.3**	**29.7**

Source: Leisure Property Services

This was valued off an 8% capitalisation rate as shown in Table 12.1.

Table 12.2 Income capitalisation, Hotel Nairobi

Net Base Cash Flow		
EBITDA in 1 values in year 1		1,001,335
Capitalised at 8.00%	12.50	12,516,692
Gross Value		12,516,692
Say		12,500,000

Source: Leisure Property Services

However, we have had to make certain special assumptions and each of these affects how much money the hotel is worth.

The first assumption we have been asked to make is a restricted marketing period for the hotel of three months. In this case, this is assumed to be too short a period to ensure that the best price is achieved. As such, the pricing of the asset will need to be reduced to ensure that a sale can be achieved within the specified time period.

The second assumption is that the accounts are not available to a purchaser. It is always worrying for a potential purchaser when the accounts are unavailable, as it could mean that something is not quite right, and in turn could deter a number of potential purchasers. As a result, to ensure a sale can be achieved without the accounts it is generally necessary to reduce the price of the asset.

The third and fourth assumptions are that the hotel is also currently closed and does not benefit from a justices' licence. These are highly onerous assumptions because the hotel will effectively have to start trading from fresh, without any pre-opening marketing, any bookings and, of course, without the benefit of a bar.

As can be seen from the Table 12.4, we have determined that the hotel will stabilise trading in three years' time, trading very poorly in the first year and improving strongly in the second year. Trading was particularly affected in the first year, when the property was only effectively open for eight months as the new owner would need time to clean up the property, recruit and train staff, as well as apply for all the relevant licenses.

The capitalisation rate adopted to value the property has been pushed out by 1% to reflect the shortened marketing period, the lack of historic trading accounts and the additional operational risk from the property.

In this example, the hotel is worth almost 25% less with the specific assumptions requested by the lending institution. However, such valuations with special assumptions are highly subjective and, as such, sometimes valuers make a spot deduction from the market value to assess the effect of the special assumptions.

The percentage deduction will obviously depend on the nature of the property, the strength of the market and the onerous nature of the assumptions. As a rule of thumb, in normal market conditions a discount is likely to be between 15% and 40%, but in exceptional circumstances there may not be a deduction or the discount could be much higher.

Valuation of Hotels for Investors

Table 12.3 Forecast in present values subject to special assumptions, Hotel Nairobi

Year	1		2		3	
No of Rooms	100		100		100	
Rooms Sold	12,775		23,725		27,375	
Rooms Available	36,500		36,500		36,500	
Occupancy (%)	35.0		65.0		75.0	
ADR	60.00		70.00		80.00	
RevPAR	21.00		45.50		60.00	
Growth (RevPAR) (%)			116.7		31.9	
Revenues ('000s)		(%)		(%)		(%)
Rooms	766.5	69.0	1,660.8	65.0	2,190.0	65.0
Food	177.7	16.0	383.3	15.0	505.4	15.0
Beverage	44.4	4.0	306.6	12.0	404.3	12.0
Total Food & Beverage	222.2	20.0	689.9	27.0	909.7	27.0
Room Hire	0.0	0.0	0.0	0.0	0.0	0.0
Leisure Club	88.9	8.0	153.3	6.0	202.2	6.0
Other Income	33.3	3.0	51.1	2.0	67.4	2.0
Total Revenue	**1,110.9**	**100.0**	**2,555.0**	**100.0**	**3,369.2**	**100.0**
Departmental Profit						
Rooms	344.9	45.0	1,079.5	65.0	1,620.6	74.0
Total Food & Beverage	44.4	20.0	207.0	30.0	363.9	40.0
Room Hire	0.0	100.0	0.0	100.0	0.0	100.0
Leisure Club	37.3	42.0	64.4	42.0	84.9	42.0
Other Income	11.7	35.0	17.9	35.0	23.6	35.0
Total Departmental Profit	438.3	39.5	1,368.7	53.6	2,093.0	62.1
Departmental Costs	672.5	60.5	1,186.3	46.4	1,276.3	37.9
Undistributed Operating Expenses						
Administrative & General	111.1	10.0	230.0	9.0	286.4	8.5
Sales & Marketing	66.7	6.0	102.2	4.0	134.8	4.0
Property Operations & Maintenance	44.4	4.0	102.2	4.0	134.8	4.0
Utility Costs	35.5	3.2	81.8	3.2	107.8	3.2
Total Undistributed Expenses	257.7	23.2	516.1	20.2	663.7	19.7
Income Before Fixed Costs	**180.6**	**16.3**	**852.6**	**33.4**	**1,429.2**	**42.4**
Fixed Costs						
Reserve for Renewals	44.4	4.0	102.2	4.0	134.8	4.0
Property Taxes	52.2	4.7	120.1	4.7	158.4	4.7
Insurance	11.1	1.0	25.6	1.0	33.7	1.0
Management Fees	33.3	3.0	76.7	3.0	101.1	3.0
Total Fixed Costs	141.1	12.7	324.5	12.7	427.9	12.7
EBITDA	**39.5**	**3.6**	**528.1**	**20.7**	**1,001.3**	**29.7**

Source: Leisure Property Services

Table 12.4 Income capitalisation subject to special assumptions, Hotel Nairobi

Net Base Cash Flow			
EBITDA in 1 values in year 3		1,001,335	
Capitalised at 9.00%		11.11	11,125,949
Less	Income Shortfall		1,435,000
Gross Value			9,690,943
Say			9,700,000

Source: Leisure Property Services

Summary

As part of the requirements of the *Red Book*, for valuations for secured lending purposes on hotels (and other properties where the value is directly linked to the trading potential of that property) the valuer has a duty to provide a "notional" valuation based upon certain artificial assumptions. These assumptions will change from instruction to instruction. Although such valuations can rarely be tested in the market, with the correct methodology it should be possible to provide the required valuations.

IFRS 15 Apportionments

Introduction

When running a business, it is usual for companies to depreciate their assets over a period of time to reflect their wear and tear and the need to purchase replacements when such assets become obsolete.

In the UK, property that is occupied by the company must also be depreciated in the company accounts and the rules governing such depreciation are outlined in UK IFRS 15.

What is depreciation?

The International Financial Reporting Standards (IFRS) states that all tangible fixed assets (other than land) will be depreciated over its/their useful economic life. The only exception being where the amount of the depreciation (both for the year and on a cumulative basis) would be insignificant. In the past, many hotel-owning companies have not depreciated buildings on the grounds that they maintain the hotel (through maintenance expenditure and periodic updates) to such a standard that makes depreciation immaterial.

The IFRS specifically states that subsequent expenditure of this type does not negate the need for depreciation and, accordingly, the Guidance Notes in the *Red Book* assume that buildings and other tangible fixed assets (except the land itself) will be depreciated.

Non-depreciation requires justifiable grounds by reference to the assets' long life and/or high residual value.

The Guidance Notes recommend that the total cost or value of a building should be divided into two categories: the "Building Core" and "Building Surface Finishes and Services".

The Building Core will typically comprise the substructure, structure, envelope and cellular completion of the building. It will have a long technical life that is not subject to periodic replacement other than for reconfiguration of the building or technical failure.

The elements of the building which are typically exposed to guests and which define the style and character of the hotel are classified as Building Surface Finishes and Services. Although these may be formed from hard wearing materials, they are likely to suffer from commercial obsolescence (ie, the trading performance of the hotel will be enhanced significantly by their replacement).

The Guidance Notes suggest that Building Surface Finishes and Services should normally be depreciated over useful economic lives of 20–30 years.

For newly constructed buildings it will generally be possible to obtain an accurate split of the total development cost between the Core and Surface Finishes and Services. For older buildings where this split is not available, an estimate should be made after consulting a quantity surveyor (probably using a deflated depreciated replacement cost of the Building Surface Finishes and Services as the starting point).

Plant and machinery and furniture and equipment will usually be carried at historical cost and depreciated over their useful economic lives.

When applying the FRS to existing hotels the valuer needs to take a view on the remaining useful economic life of each property, its residual value and the allocation between the various asset components that make up the total property.

In its executive summary on tangible fixed assets for the hospitality industry, the British Association of Hospitality Accountants (BAHA) produced a briefing paper on FRS 15. Below we outline some of the salient points for the hotel valuer.

Guidance Notes have been prepared in response to Financial Reporting Standard (FRS 15) "Tangible Fixed Assets" issued by the Accounting Standards Board (ASB) in February 1999. Following the publication of the ASB's proposals for the initial measurement, valuation and depreciation of tangible fixed assets in its October 1996 Discussion Paper, followed by Financial Reporting Exposure Draft (FRED) 17 in October 1997, a committee of accountants was formed under the auspices of the British Association of Hospitality Accountants (BAHA) representing a broad cross section of owners and operators of hotels, hotel accountants and independent auditors. Subsequently, discussions were held with some of the leading Valuers of hotels.

The objective of this committee was to review accounting standards and practices and formulate a set of guidance notes which will facilitate a uniform framework and ensure that there is more consistency in accounting for fixed assets within the industry.

These Guidance Notes are not intended to be prescriptive, but to set out an approach generally accepted by the industry and provide guidance for preparers of hotel company accounts. The Notes are intended to be indicative of best practice.

Consistent with Financial Reporting Standards, the committee endorses the objectives that:

"consistent principles are applied to the initial measurement of tangible fixed assets; where an entity chooses to revalue tangible fixed assets, the valuation is performed on a consistent basis, kept up-to-date and gains and losses on revaluation are recognised on a consistent basis; depreciation of tangible fixed assets is calculated in a consistent manner and recognised as the economic benefits are consumed over the assets' useful economic lives."

Initial measurement
In accordance with general requirements, hotel properties and other tangible fixed assets should initially be recorded at cost. Any capitalisation of costs incurred in bringing the hotel into use should include only directly attributable costs and should cease once the hotel is substantially complete, even if it has not yet been brought into use.

If the cost of a new hotel exceeds its recoverable amount, it should be written down to its recoverable amount. This is likely to be the case only if the performance of the hotel is materially worse than was anticipated in its feasibility appraisal, or if there are material overruns in the cost of construction, or if the company overpaid for it by a significant amount.

Subsequent expenditure that ensures the hotel maintains its overall standard of performance should be charged to the profit and loss account as incurred. Subsequent expenditure should be capitalised only when it provides for new or enhanced revenue streams (for example, the addition of a leisure facility or the upgrading of conference rooms to a standard that enables significant new business to be attracted), or it relates to the replacement or restoration of fixed assets which have been treated separately for depreciation purposes and depreciated over their useful economic lives.

It is recognised that on implementation of FRS 15 many hotel companies will not have fixed asset records that enable them to separately identify assets in sufficient detail for these purposes. The Guidance Notes suggest that, in these circumstances, approximations of the cost and accumulated depreciation of the assets will need to be made in order to ensure that these amounts are eliminated from the balance sheet when the assets are replaced.

Valuation

The Guidance Notes envisage that many hotel-owning companies will revalue their properties periodically, although it is recognised that some may choose to record them at historical cost or retain the book amount that reflects previous valuations. Where revaluation is adopted, in accordance with the requirements of the FRS, a full external valuation should be carried out at least every five years with an interim valuation (either external or internal) in the third year and annual reviews for impairment in other years where there is an indication that impairment may have occurred. Alternatively, the portfolio may be valued on a rolling basis such that all properties are revalued at least once every five years.

Hotels are classified as non-specialised properties and should generally be valued on the basis of existing use value, unless they are surplus to requirements in which case open market value should be used.

A hotel valuation for balance sheet purposes will normally involve a valuation of the entire business, on the basis of existing use as an operating hotel, inclusive of trade furniture, fittings and equipment. The allocation of the overall value between land, buildings and other items (including plant and machinery, motor vehicles, furniture and fittings) should be carried out by the Valuer in accordance with the RICS Red Book (GN5). Generally, allocations will be estimated for the elements other than land based on depreciated replacement cost, with the balance allocated to land. There will be instances where it is more appropriate to allocate some of the residual balance to the building. However, the allocation should produce results that appear reasonable to the Valuer in all the circumstances.

The *Red Book* is quite clear on how these need to be calculated, the detail of which is outlined in practice statement 1.8.

The key points for the hotel valuer to remember are:

- The basis of value will usually be existing use value.
- This value must be apportioned between Land, Buildings and Fixtures & Fittings. (It is not appropriate for trading assets to split the trading potential out as it is an intrinsic part of the value of the hotel).

Example 43: Hotel Havana

In this example, the existing use value has been determined at £14,000,000.

Table 13.1 Existing use value, Hotel Havana

Bedrooms	120	
EUV	14,000,000	
Apportionment	(£)	(%)
Land	3,920,000	28.00
Building	8,960,000	64.00
Building — Core	6,216,000	44.40
Building — Services & Finishes	2,184,000	15.60
Plant & Machinery	560,000	4.00
FF&E	1,050,000	7.50
Computers	70,000	0.50
Total	14,000,000	100

Source: Leisure Property Services

The land value has been arrived at by referring to comparable sales transactions, which in this instance show land values at £32,000 per bedroom for hotels of this size.

The FF&E allowance (including computers) for a hotel of this type come to just under £9,500 per bedroom, and after cross-referring with the building surveyors it has been agreed that the structure and internal finishes of the property would be in the region of £75,000 per bedroom.

If the property had only just been built, it would be normal to add more detailed analysis of the building costs, which may allow for a more accurate breakdown between the elements of the construction.

Summary

The International Financial Reporting Standards require that all tangible fixed assets (except for land) be depreciated over the period of their anticipated useful life. Whether regular FF&E allowances are used to keep the properties economically useful is immaterial to this requirement. To comply with these requirements, hotel valuers are required to artificially apportion value between the different parts of a hotel. Although not a formal valuation, such apportionments require no less stringent a standard of duty and, as such, it is essential that the valuer follows the relevant local professional guidelines when carrying out such assessments.

Concluding Summary

Working out the value of a hotel requires the valuer to put himself in the position of the potential purchaser to assess how desirable that particular property is. Therefore, it is essential that the valuer looks at all the same factors that a buyer would look at, including what lies behind a buyer's motivations what the true sustainable profit at the hotel is likely to be, and being aware of the specific local market conditions and the macroeconomic outlook.

It is also essential that a valuer follow all the professional guidance provided by the various professional associations, with certain valuations requiring specific and detailed steps to be followed to ensure that the valuation advice is accurate and complies with such requirements.

This book should have provided the basic first steps for a valuer to understanding how to approach a hotel valuation, as well as helping the potential purchaser to understand the pricing mechanisms of the hotel sector, and explain some of the technical aspects behind valuation to hoteliers and financiers. In the earlier chapters, the driving forces behind hotel ownership are briefly explored, outlining some of the principles behind the "buying decision", as well as looking more closely at the fundamentals behind the hotel investment market. Understanding how to assess the trading potential behind a particular hotel and exploring the local market that it sits in is outlined at some length and this should enable potential purchasers and valuers to approach the problem of assessing the real potential of a hotel.

The hotel examples within the book should have provided guidance on how valuations are carried out in each circumstance, from rent reviews through to purchases of leasehold investments, including IFRS 15 taxation apportionments through to site valuations.

It should be noted, however, that hotel valuation is a detailed subject and requires specialist knowledge; it will be difficult to undertake accurate hotel valuations without fully understanding the dynamics behind the hotel market, the active purchasers in the particular sector of the market, and the prevalent sentiments for the likely future trading performance in that particular area. Hopefully, this book will have helped to guide the novice through some of the trickier areas and will enable both owners and advisers to know where they need specialist advice.

Glossary

Below we have listed a wide range of definitions that are used throughout both the hotel industry and in valuations.

ADBUs	Adjusted double bedded units.
ADR	Average daily rate (sometimes known as average room rate). This is the total rooms' revenue (over whatever period) divided by the number of rooms occupied. Sometimes this is calculated by only taking into account paid rooms rather than including complimentary rooms.
ARR	Average room rate (also referred to as the ADR). See ADR for fuller details.
Average food check	This is a calculation that works out the average spend per visit in the restaurant /bar. It can be calculated many ways but in essence it is a simple tool to show whether the level of expenditure per guest is increasing or decreasing.
A&G	Administration and general.
Capex	Capital expenditure (capex) — funds used by an organisation to acquire or upgrade physical assets, such as property or equipment.
C&B	Conference and banqueting.
DBUs	Double bedded units.
Depreciation	The measure of the cost, or revalued amount, of the economic benefits of a tangible fixed asset that have been consumed during the period. Consumption includes the wearing out, using up or other reduction in the useful economic life of the tangible fixed asset whether from use, effluxion of time or obsolescence through either changes in technology or demand for goods and services produced by the asset.
DV	District valuer.

EBIT	Earnings before interest and taxation.
EBITDA	Earnings before interest, taxation, depreciation of amortization (sometimes known as net operating profit).
EDBUs	Equivalent double bedded units.
EHO	Environmental health officer.
External valuer	A valuer who, together with any associates, has no material links with the client, an agent acting on behalf of the client, or the subject of the assignment.
FMR	Fair maintainable receipts.
Fixed costs	These are expenses below the IBFC line that need to be deducted to calculate the EBITDA or net operating profit of the hotel. They typically include property taxes, insurance, rent, an FF&E allowance, management costs, non-departmental leasing costs and franchise fees.
F&B	Food & beverage department. This is a specific cost centre for hotels and generally includes all revenue generated from food and beverage sales (as would be expected), including revenue from the restaurant, bar, room service and conference and banqueting.
	F&B revenue is also sometimes broken down and reported as Food, Beverage and Other F&B revenue and departmental profit streams.
FF&E reserve	Furniture, fittings and equipment (FF&E) reserve. This is atypically a percentage of turnover that is deducted each year as a sinking fund to enable the hotel to ensure the public areas are regularly maintained.
Financial statements	Written statements of the financial position of a person or a corporate entity, and formal financial records of prescribed content and form. These are published to provide information to a wide variety of unspecified third-party users. Financial statements carry a measure of public accountability that is developed within a regulatory framework of accounting standards and the law.
GOP	Gross operating profit (sometimes also known as IBFC).
GOPAR	Gross operating profit per available room. This is a measure being introduced into the hotel industry to try and calculate profitability ratios of hotels. It is calculated by taking the GOP of the hotel and dividing it by the number of bedrooms.
GTAS (gross takings all sources)	The valuation office's term for total revenue.
IBFC	Income before fixed costs (sometimes referred to as gross operating profit).
Independent valuer	A valuer who meets the specific requirements of independence, prescribed by law or regulation, for particular valuation tasks in certain States.
Internal rate of return (IRR)	The rate of interest at which all future cash flows must be discounted in order that the net present value of those cash flows, including the initial investment, should be equal to zero.

Loan to value (LTV)	Loan to value is the proportion a bank will lend on an asset in relation to its overall market value.
Market rent (MR)	The estimated amount for which a property, or space within a property, should lease on the date of valuation between a willing lessor and a willing lessee on appropriate lease terms, in an arm's length transaction, after proper marketing wherein the parties had each acted knowledgeably, prudently and without compulsion.
Market value (MV)	The estimated amount for which a property should exchange on the date of valuation between a willing buyer and a willing seller in an arm's-length transaction after proper marketing wherein the parties had each acted knowledgeably, prudently and without compulsion.
Market value with existing use	The estimated amount for which a property should exchange on the date of valuation between a willing buyer and a willing seller in an arm's-length transaction, after proper marketing wherein the parties had acted knowledgeably, prudently and without compulsion, assuming that the buyer is granted vacant possession of all parts of the property required by the business and disregarding potential alternative uses and any other characteristics of the property that would cause its market value to differ from that needed to replace the remaining service potential at least cost.
Mortgage lending values (MLV)	Mortgage lending value shall mean the value of the property as determined by a valuer making a prudent assessment of the future marketability of the property by taking into account the long-term sustainable aspects of the property, the normal and local market conditions, as well as the current use and alternative possible uses of the property.
M&E	Meetings and events.
Net initial yield (NIY)	The net initial yield is the initial net income at the date of purchase expressed as a percentage of the gross purchase price including the costs of purchase.
Net operating profit (NOP)	Net operating profit (sometimes also referred to as EBITDA).
Occ	Occ is a shorthand abbreviation for the occupancy rate at the hotel. This is calculated by multiplying the number of used bedrooms (over whatever period) by the number of available rooms. So, for example, if one hotel had 30 bedrooms and had over a period of one week sold 184 rooms that would equate to an occupancy rate of 86.66% ($184/(7 \times 30)$). However, if one of the rooms was unavailable, then the occupancy rate would be 90.64% ($184/(7 \times 29)$).
Personal goodwill	Personal goodwill is defined as 'the value of profit generated over and above market expectations which would be extinguished upon sale of the specialised trading property, together with those financial factors related specifically to the current operator of the business, such as taxation, depreciation policy, borrowing costs and the capital invested in the business' (IVS GN 12).
POM	Property operations and maintenance.

POMEC	Property Operations & Maintenance & Energy Costs.
P&L	Profit and loss accounts.
Reasonably efficient operator (REO)	A reasonably efficient operator is defined as 'a market-based concept whereby a potential purchaser, and thus the valuer, estimates the maintainable level of trade and future profitability that can be achieved by a competent operator of a business conducted on the premises, acting in an efficient manner. The concept involves the trading potential rather than the actual level of trade under the existing ownership so it excludes personal goodwill'.
Red Book	*The Valuation & Appraisal Manual*, a book that provides professional guidance for all chartered surveyors and valuers.
RevPAR	Revenue per available room. This is one of the standard benchmarking measures in the hotel industry. However it is not straight forward as the revenue related to rooms' generated revenue rather than total revenue. It is calculated by taking all rooms revenue and dividing by the number of bedrooms. It can also be calculated by multiplying the ADR (or ARR) by the occupancy rate.
RICS	Royal Institution of Chartered Surveyors.
R&M	Repairs and maintenance.
RV	Rateable value.
Specialised trading property	Property with trading potential, such as hotels, fuel stations, restaurants, or the like, the market value of which may include assets other than land and buildings alone. These properties are commonly sold in the market as operating assets and with regard to their trading potential. Also called property with trading potential.
Years purchase (YP)	Years purchase. This is the number that the stabilised profit (or income for an investor) is multiplied by to arrive at the purchase price (or how many years it would take for the original investment to pay off the purchase price in very simplistic terms.) So, for example, if a hotel is generating £190,000 stabilised profit and it is sold for £2,400,000 that would equate to a YP of 7.9 (190,000/2,400,000). This, in turn, can demonstrate the capitisation rate used. In this case, (100/7.9) it represents a capitalisation rate of 12.66%.

Index

Airport hotel 7
Alternative uses 20, 45, 69, 71, 73, 74, 75, 83, 123
Apartotel. 10

Basle II. 70, 71
Benchmarking 22, 25, 27, 133, 136
Bijou hotel. 6
Boutique hotel 6, 122
Budget hotel 9, 18, 19, 20, 22, 23, 156

Conflicts of interest. 67

Deloittes 27
Depreciation 11, 16, 28, 30, 31, 59, 61, 63, 71, 81, 82, 84, 93, 89, 185, 186, 187

Five-star hotels. 5, 7, 10, 22, 122

Ground leases. 38, 129, 130

Heron 52
Hotel grading. 10

Land Securities 52
Leisure Property Services 28, 30, 32, 33, 34, 36, 37, 92, 93, 94, 96, 97, 98, 99, 100, 101, 103, 104, 105, 107, 108, 109, 110, 111, 112, 113, 114, 117, 127, 128, 145, 146, 147, 148, 140, 141, 142, 143, 144, 158, 160, 161, 162, 154, 174, 175, 176, 177, 180, 181, 182, 183
Lifestyle hotels. 2, 3, 99
London & regional 52

195

Management contracts 4, 29, 124, 128, 129, 132, 133, 134, 135, 153
Marcol 52
Mid-market hotel 6, 22, 27, 106, 108, 122, 169

Operational risk 54, 132, 135, 181

Personal goodwill 15, 41, 59, 60, 61, 62
Property goodwill 15, 41

Seaside hotel 8
Spa hotel 7

The Bench 27
Townhouse hotel 6
Traditional building leases 130
Treasure hunting 19, 123
TRi 27

Uniform System of Accounts 27, 29, 31

Valuation Information Paper 15, 31, 74, 106

Hotels
Hotel Amman 169, 170
Hotel Ankara 154

Hotel Basseterre 125
Hotel Beijing 145, 146, 147
Hotel Belfast 126, 127
Hotel Berlin 137
Hotel Brussels 116
Hotel Bratislava 169
Hotel Bridgetown 135
Hotel Boa Vista 107, 108
Hotel Bucherest 174, 175, 176, 177, 178
Hotel Budapest 147

Hotel Cairo 136
Hotel Canary 156, 157, 158
Hotel Canberra 128
Hotel Cardiff 145, 146
Hotel Cayman 156, 157, 158

Hotel Dar Es Salaam 125
Hotel Dodecanese 156, 157, 158

Hotel Family	156, 157, 158,
Hotel Harare	172
Hotel Havana	188
Hotel Jakarta	160
Hotel Kampala	111, 112
Hotel Lilongwe	104, 105
Hotel Lisbon	140, 141
Hotel Ljubljana	163
Hotel London	161, 162
Hotel Lusaka	100, 101, 103, 105
Hotel Moscow	121
Hotel Nairobi	179, 180, 181, 182, 183
Hotel Nicosia	113, 114
Hotel Oslo	141, 142, 143
Hotel Palermo	116
Hotel Paris	107
Hotel Rabat	156
Hotel Reykjavik	96, 97, 99
Hotel Sal	107, 108
Hotel Sao Vincente	107, 108
Hotel Sarajevo	92, 93
Hotel Scarborough	135
Hotel Sofia	165
Hotel St John	102, 103
Hotel Stockholm	167
Hotel Suva	109, 110, 111
Hotel Valletta	98, 99
Hotel Victoria	127, 128
Hotel Vienna	109, 110, 111, 112, 113, 114
Hotel Vilnius	94, 95
Hotel Zagreb	143, 144

The Secret Life of Syrian Lingerie

The Secret Life of Syrian Lingerie

Intimacy and Design

By Malu Halasa and Rana Salam

Featuring text by Noura Kevorkian, an interview with Ammar Abdulhamid, poetry and self-portraiture by Iman Ibrahim, and photographs by Reine Mahfouz, Issa Touma, Gilbert Hage, and Omar Al-Moutem

CHRONICLE BOOKS
SAN FRANCISCO

Prince Claus
Fund Library
is gratefully acknowledged
for its support of this project.

Introduction and compilation copyright © 2008 by Malu Halasa and Rana Salam.
"Competing Thongs" copyright © 2008 by Malu Halasa.
"On Love and Lingerie" copyright © 2008 by Noura Kevorkian.
"A Fundamentalist Changes His Mind" copyright © 2008 by Malu Halasa.
Photographs "Damascus" copyright © 2008 by Reine Mahfouz.
Photographs "Portraits of Women" copyright © 2008 by Issa Touma.
Photographs "Up Close" copyright © 2008 by Gilbert Hage.
Photographs "Modeling Lingerie" by Omar Al-Moutem copyright © 2008 Rana Salam Design.
Photographs and text "Coda: A Room of One's Own" copyright © 2008 by Iman Ibrahim.

All rights reserved. No part of this book may be reproduced in any form
without written permission from the publisher.

Library of Congress Cataloging-in-Publication Data available.

ISBN: 978-0-8118-6458-9

Manufactured in Hong Kong
Design by Brett MacFadden

Financial support for this publication was provided by
Prince Claus Fund for Culture and Development
Hoge Nieuwstraat 30
2514 EL The Hague
The Netherlands
www.princeclausfund.org

10 9 8 7 6 5 4 3 2

Chronicle Books LLC
680 Second Street
San Francisco, CA 94107
www.chroniclebooks.com

Table of Contents

Introduction	7
Damascus: From Factory to Souk Photographs by Reine Mahfouz	11
Competing Thongs: The Lingerie Culture of Syria Essay by Malu Halasa	
Portraits of Women Photographs by Issa Touma	35
On Love and Lingerie Essay by Noura Kevorkian	
A Fundamentalist Changes His Mind: Sexuality and Humor in Syria Interview with Ammar Abdulhamid	57
Up Close: Intimate Still Lifes Photographs by Gilbert Hage	69
Inside Story Interviews with Syrian Women	
Modeling Lingerie: Product Photography from Lingerie Manufacturers Photographs by Omar al-Moutem, Amasel Fashion, and Angel Lady	127
Coda: A Room of One's Own Poetry and Self-Portraiture by Iman Ibrahim	155
Footnotes	166
Bibliography	168
Editors and Contributors	170
Index	173
Acknowledgments	176

Introduction

When we started trawling the Syrian lingerie shops together in 2003, some of Rana's favorite styles were still in demand. Others had disappeared in the blink of a season. No G-strings featuring noisy rubber insects were displayed in the women's clothing store we visited. We did find four floors of everything from embroidered dresses, bathrobes, sheer negligees, and feathered bra-and-thong sets to lavish polyester wedding gowns filling an entire floor. After salesmen spent hours rooting around—more than likely in Souk al-Hamidiyeh's nearby stalls and shops—they presented us the next day with an incandescent cockroach on a thong. Among the other items we purchased was a pair of silver-trimmed, white platform wedding shoes more suited to a 1970s transvestite disco than a walk down the aisle.

Syrian design is schizophrenic. Ages and influences compete with each other. According to Syrian political commentator and novelist Ammar Abdulhamid (interviewed for this book), the country is ancient and postmodern at the same time. "If you take a cross section of Syrian culture, you are going to see a spectrum of different cultural values that cross a thousand years, but they all exist right now in one single moment," he says.

Lingerie is no different. Styles zigzag from prim virginal floral arrangements crowning a thong like a wedding corsage to nippleless leotards reminiscent of Frederick's of Hollywood. There are colorful plastic butterflies and flowers sewn onto underwire bras and zippered breasts and crotches verging on a crudely innocent version of S&M. Some of the bra-and-panty sets sing and light up. Others can be eaten.

Further investigations reveal that lingerie follows a twice-yearly fashion cycle. The companies design new styles and photograph them for the buyers who come to the Motex clothing and trade exhibition to place advance orders. Manufacturers jealously protect their best-selling styles, and deep-seated rivalries between companies are not uncommon. Designers, photographers, and models fuel a fashion industry little known outside the world of the people who make the lingerie and those who buy it.

Globalization and the Internet mean Syrian lingerie design did not develop in a vacuum. It has been shaped by a unique set of social conventions. In the West, sex sells lingerie and pretty much everything else. Syria is a deeply religious and conservative society slowly emerging from a thirty-year, Pan-Arab socialist dictatorship. Lingerie, a luxury item, developed out of the economic boom that followed the 1973 Yom Kippur War with Israel. It is part of the popular working-class tradition surrounding marriage for the country's mainly Muslim population. Symbolically, lingerie represents an important rite of passage from virginity to respectable, married womanhood.

The country's stringent censorship laws prevent the publication of revealing images of women, yet photographs of Eastern European women modeling lingerie openly circulate in the souk. In 2003 many were kept in hand-assembled catalogs provided to the stores and stall owners by the lingerie companies. Two sixtysomething gentlemen behind a wooden stand—every inch of which was covered in panties and bras—thought nothing of showing us a thick red photo album with page after page of sweetly smiling models. With its openings and "grab holes," the lingerie is obviously about sex, but its public face is at best innocent and unthreatening.

But souk and catalogs tell only part of the story. The factories we were finally able to visit two years later in 2005 revealed other deeply rooted motivations. Reine Mahfouz's fly-on-the-wall photographs of factory rooms in the basements of residential buildings in Damascus, and of wholesale showrooms, provide a wider context for both the lingerie and the society that produces it.

In spite of U.S. sanctions and a world market increasingly dominated by cheap textiles from China and India, the lingerie industry in Syria is growing. Boxes of crocheted bodysuits and chocolate hearts featuring G-strings are just a few of the sex-related products and clothing exported across the Middle East.

Through lingerie design, the more elusive subject of intimacy can be approached at a time when there are many misconceptions about the Arab world because of gender segregation and veiling. On the surface, Syrian lingerie is about women's wants and desires. However, in an Islamic culture where core spiritual beliefs inform the status of women, and where men control all aspects of production, lingerie addresses both the use and the abuse of the female body as well as deep-seated, streetwise religious attitudes about the purpose of women in the service of their husbands. Syrian women have their own ways of confronting authority, and they confide in filmmaker Noura Kevorkian, who recorded their thoughts and confidences in a journal kept during the filming of her 2001 documentary *Veils Uncovered*.

Because of war, assassination, and strife, Syria remains a country in political turmoil. The meteoric rise of its capital, Damascus, was a direct result of Beirut's demise during the Lebanese civil war. At a time when conservative religious scholars in Egypt issue fatwas against nudity during marital sex,[1] Islamic society in Syria is not demure or prudish. The Syrians are known for being blunt and forthright in matters sexual, as seen by some of the country's popular jokes and sexual innuendos. At women-only parties, given for a bride before her marriage night, women sometimes dress in revealing clothing. Flesh, taboo in public, is flaunted in private.

In cultures where women are veiled and/or segregated, flamboyant outerwear is not meant for public consumption. Bright colors, ornateness, and kitsch occupy interiors, imaginative or otherwise. This playfulness is accentuated in still-life photographs by the Lebanese artist and photographer Gilbert Hage. Does the lingerie belong in a museum or the boudoir? At this book's secret heart is the self-portraiture and poetry of Iman Ibrahim, working from her bedroom in Aleppo, Syria's deeply religious second city, home to the Muslim Brotherhood.

Popular design is often shrouded by the region's magnificent past. After the Muslim conquest, the people of Arabia emerged from war to create an opulent culture. To paraphrase historian Philip K. Hitti, they went from the desert and scorpions to palaces and sherbets. Lingerie is a continuum of an ancient trade. Through its creation and design, a people's story can be told. The engagement of the arts and society acts as an invaluable prism onto a place at a time when much of the writing about the Arab world repeats the weary rhetoric of war.

The Secret Life of Syrian Lingerie: Intimacy and Design explores a nation's public image as reflected in its textile industry, sexual politics, and religion. The modern Middle East is glimpsed through one of the region's more opaque glasses: Syria. The vantage point of people working inside the country's thriving lingerie-manufacturing culture provides a view of the distinct intimate space where men and women have the possibility of meeting as equals.

Malu Halasa and Rana Salam, London

Photographs by Reine Mahfouz

Damascus: From Factory to Souk

Reine Mahfouz is one of the few women photographing arts and culture in Beirut. The reconstructed war-torn city, its changing political history, and its cityscape, have been a main subject. In an artist's statement, "Angel Lady," for the 2007 Paris exhibition *Sexy Souks*, on Middle East eroticism,[1] she writes, "Amidst the racket of the city I try to understand the structure of the space where we live, and the thickness and the depth of the layers upon which it was built." She brings the same excavation skills to Syrian lingerie. She continues, "In order to establish the brand 'Made in Syria,' Syrian lingerie manufacturers nourish the most extravagant sexual fantasies in the strange quaintness of the lingerie. In my photographs, I capture the mix of religion and intimacy, the political with the erotic. It is an optical exploration that questions our prejudices surrounding the most deep-seated customs in the evolution of Arab society."

Essay by Malu Halasa

Competing Thongs: The Lingerie Culture of Syria

1

The Syrians have taken their former dictator's advice to heart. Hafez Assad warned them that the world would one day turn its back on them and they should be completely self-sufficient. That time is now.

Faced with growing international criticism and the effects of U.S. sanctions, Syrians and their manufacturing remain resilient. The Syrian cotton industry, ranked eleventh in the world, has been a magnet for companies like Benetton, Naf Naf, and Stefanel in a modern continuation of one of the earliest professions—the trading of textiles and clothing. Damascus is considered the oldest inhabited city in the world; the Uzbek pilgrims, who still stop there on their way to Mecca, pray in the city's Umayyad Mosque, a place of worship for millennia, and exchange their hand-embroidered Suzani bedspreads and tablecloths for paper-thin Syrian terrycloth—the only bearable wrap in the baking heat of Saudi Arabia.

The Secret Life of Syrian Lingerie

Above: In underwear alley of Souk al-Hamidiyeh in Damascus, men, women, and children shop, work, and relax in the shadow of skimpy negligees.

Page 10: One of the few women who design Syrian lingerie, Sumaya al-Ali prepares a rush order of Lingerie Lour's crochet one-piece body suits, bra-and-panty sets, and hip shawls for a hairdresser's salon in Tripoli.

Photographs by Reine Mahfouz   Essay by Malu Halasa

The Syrians, known as the Chinese of the Middle East, import relatively little and manufacture all manner of clothing themselves before exporting it abroad, from chadors for Iran to racy lingerie destined for the malls of Saudi Arabia.

In the eighteenth-century bazaar of the Old City, Souk al-Hamidiyeh, it is not uncommon to see veiled Shi'i and Sunni Muslim women in long-sleeved, floor-length coats appraising window displays featuring feathered bras and panties and one-piece crocheted bodysuits with "grab holes" at nipple-level alongside traditional dresses and over-the-top white polyester wedding gowns sprinkled with sequins, diamantés, and lace. Sunni Syrians love white weddings and sexy lingerie, which hangs in elaborate displays on torso mannequins or torso-shaped hangers from the eaves of underwear stores and from stalls. Lingerie has become an essential part of the wedding trousseau. If a groom doesn't buy the lingerie for his wife-to-be, the bride herself or her mother does, sometimes collecting up to thirty outfits for her wedding night. The fashion scene is thriving and competitive, but it is lingerie design that reveals an unexpected Syrian frankness and bawdy sense of humor toward sex. It feeds into prurient interests in both the East and the West about what lies beneath the veil, and also tells a unique story about fashion, dictatorship, and sex tourism.

In underwear alley off Souk al-Hamidiyeh's main thoroughfare, a glass counter fills the tiny hole-in-the-wall shop Lingerie Baqdounis ("parsley" in Arabic). Floor-to-ceiling shelves are stuffed with thin white cardboard lingerie boxes. Owner Hamas Baqdounis sells women's intimate apparel—traditional bras and panties—but his specialty is the humorous variety popular among newly engaged Muslim couples.

The first thongs he shows are adorned with colorful fake birds: *ish al-asfour* (bird's nest) is slang for women's pubic hair. He follows them up with feathered eggs, each containing a surprise inside. Proving his point, Baqdounis rummages around for a cracked egg. Barely visible through the broken shell is the thinnest red silk sliver of a G-string. Nests and eggs give way to colorful flowered bras and thongs in garish oranges, reds, and greens, and Lingerie Parsley blooms into a garden of Paradise. Some of the lingerie lights up, and one orange daisy bra, when pressed, sings "Old MacDonald Had a Farm." Syrian lingerie draws from the visual traditions of the Arab street—kitsch. The rose of martyrdom, a symbol of Hizbullah, is made of red cloth, kissed by plastic dewdrops, and is glued to a thong. But kitsch reflects the popular and the widely available, and like everywhere else in the world, Syria has been invaded by cheap electronic goods, toys, and cloth from China.

The pièce de résistance at Lingerie Baqdounis is a front-buckling gold lamé bra with matching panties and a micro-miniskirt. "This is for the woman who dances the Lambada for her husband." The proprietor's eyes glint mischievously. "If the first wife doesn't dance the Lambada, he will divorce her and get a second one. If the second wife doesn't dance the Lambada, he will divorce her and get a third one. By the fourth wife he will have found a woman who dances the Lambada." (The Syrian-style Lambada was a popular bedroom pole-dance in 2003, the year Rana Salam and I started shopping together for lingerie.)

Two years later, a new lingerie style was gracing the bedroom: plump red velveteen hearts, one on each breast and on the crotch, stenciled with the words *ana behibik* (I love you) in white, as if done with icing. When squeezed, the hearts play a little song. The lingerie is *sha'abi*, popular verging on vulgar, but people tire of the same old jokes, and manufacturers are under enormous pressure to create new designs or duplicate popular models. As one divorcée admitted, she changed into different sets of lingerie during the first week of marriage, including novelty thongs for breakfast.

In al-Adnan, a lingerie shop on the main street of the souk, an Iranian Shi'i woman in an *abaya* shops for lingerie. She examines a bra-and-thong set made from material that resembles faux coconut skin—too scratchy. She sniffs an edible thong and makes a face—too smelly. The style that catches her attention is a fluorescent green bra-and-thong set complete with arm and leg bands—all silver fringed—manufactured by Amasel Fashion. She peers at a photograph, showing a non-Syrian woman modeling the style, attached to the outfit. Her husband and teenage son wait by the door.

Another woman, a Syrian Sunni in her early thirties, with a headscarf clipped under her chin, enters the shop. She is with an older sister and elderly father who remain outside. Her makeup is distinctive: heavy eyeliner and two shades of lipstick. Soon to marry, she smiles demurely before saying she has no "imagination" when it comes to her wedding night. She would like to buy an assortment of lingerie, but al-Adnan is notorious for not giving discounts. So she satisfies herself with a foil-covered chocolate heart for 100 Syrian pounds (approximately $3).

Muhammad Emad Haliby, thirty, manages a women's clothing store and shops regularly at al-Adnan. "There has to be laughter in the bed," he says. "If the man doesn't make the woman laugh, the sex is dead. The more she laughs, the better the sex. If Syrian lingerie didn't exist, sex would be boring."

Photographs by Reine Mahfouz   Essay by Malu Halasa

Above: Owned by the Mousali and Omari families (whose names appear on the sign), the hole-in-the-wall shop Al Batul specializes in a wide range of lingerie and belly-dancing outfits in Souk al-Hamidiyeh.

Page 17: Wholesale customers select styles off the racks or from Amasel Fashion's hand-assembled photo albums featuring the latest creations by the firm's proprietor, Mouhammed Doukmak.

Edible bikinis—made from a waxy Nescafé-flavored paste—were a popular new style when his wife first wore them in 2003. She keeps a cupboard of lingerie, and in the morning before Haliby goes to work, he sometimes chooses an outfit and leaves it on the bed. Other times he calls from the office and asks, "What's cooking tonight?" She is not the only one expected to dress up for their appointment. He wears satin shorts and cologne.

As a teenager, Haliby observed lingerie from afar but never discussed it. He waited until marriage. For him, the lingerie represents sexual freedom and safe sex. "As long as you get everything at home, you don't get sidetracked and go to prostitutes."

Haliby nonchalantly leafs through the pictures of Eastern European women modeling the latest lingerie styles in the album on al-Adnan's counter. There is a good chance he's looking at prostitutes. In each image, a woman in a sexy maid's costume or a latex leotard with lewd cutouts stands in front of a hand-painted backdrop such as a sylvan glade or a Chinese pagoda and fountain. With their hands on their hips, or holding a duster, many smile sweetly as if posing on holiday. The six-by-four-inch photographs in the albums or attached to lingerie outfits enable prospective customers—both women and men—to see what they're buying.

One photo has the telephone number of thirty-four-year-old photographer Omar al-Moutem stamped on the back. In 2003, al-Moutem turned the front room of his house near the souk into a studio with three backdrops, including a log cabin. From the beginning, he spoke openly about lingerie and sex. "Although there are a lot of openings"—he was referring to the grab holes and crotchless panties—"these are not used for anal sex." As a good Muslim, he felt he should be specific. For him, religion and lingerie are compatible as long as his pictures aren't titillating. Despite his candor, which surprised Rana (who is from a Lebanese Sunni family), al-Moutem's new wife served cold drinks through an opened doorway, with only her hands visible.

When we return in 2005, the photographer explains that the secret of Syrian lingerie lies in the souk. He takes us back through the narrow alleyways behind the city's thirteenth-century citadel—the first of many trips together around the city. Dar al-Hadith al-Nuriya Lane 0302 overflows with all kinds of plastic paraphernalia from China—beeping kids' mobile phones, rubber insects with beady red eyes for attaching to thongs, and ornamental bottles and sprayers for holding the latest edible sex creams and massage oils. There are fake flowers

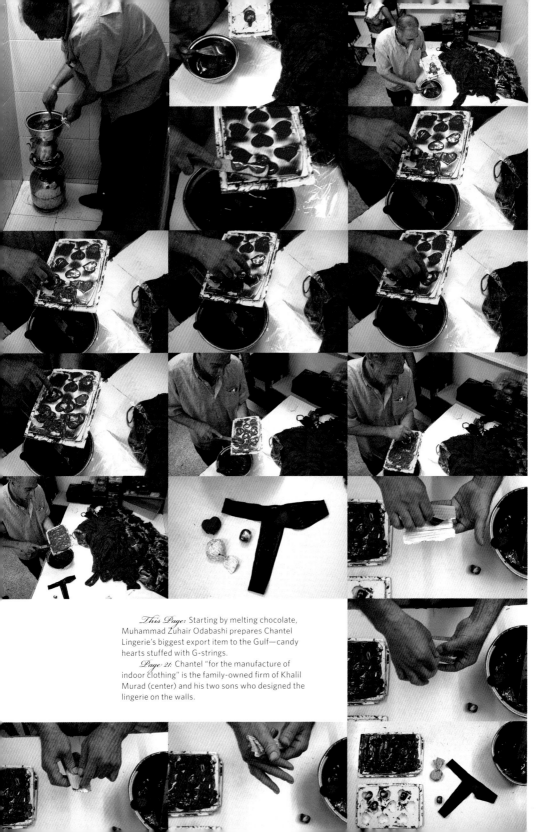

This Page: Starting by melting chocolate, Muhammad Zuhair Odabashi prepares Chantel Lingerie's biggest export item to the Gulf—candy hearts stuffed with G-strings.

Page 21: Chantel "for the manufacture of indoor clothing" is the family-owned firm of Khalil Murad (center) and his two sons who designed the lingerie on the walls.

everywhere. The new China has reached the Old City, and the Damascenes, long used to foreign invasions, react instinctively. They adapt and adopt, refashion and embellish, and sell on again. Stifled by the outside world as well as by the actions of their own government, the Syrians survive on their wits.

2

Beyond the timeless walls of the Old City rises a capital of dreary cinderblock and stone. Damascus, once filled with traditional houses and courtyards, fell under developers' wrecking balls in the late 1970s. Now the city feels unfinished and unloved. There is a failing infrastructure, and the capital once known as *al-Fayha* (the Fragrant) is filled with the festering smell of the Barada River. The streets are dry, dusty, and dirty, and there are people everywhere during the rush hour for the main midafternoon meal.

Chantel Lingerie "for the manufacturing of indoor clothing" operates in the anonymous basement of a residential building near Baghdad Street, at the bottom of broken stone steps, past a refrigerator on the landing. The "indoor clothing" hangs on the wall behind the sewing machines: a red boa–trimmed pantsuit, baby-doll nighties, and floor-length, see-through white negligees. Next door, a mannequin sports a sheer veil plus a top and pantaloons decorated with silver sequins—a style still under construction.

At Chantel, sex candy sells well. The firm of Khalil Murad and his two sons manufactures pungently flavored chocolate, mango, strawberry, and tutti-frutti edible bikini sets. In a culture where religion prevents men and women from meeting alone, other signals apply. A G-string in a heart-shaped chocolate or glacé fruit makes intentions crystal clear. Depending on the demand from Saudi Arabia and Kuwait, Damascene lingerie companies manufacture an estimated 20,000 chocolate hearts every month, so maintains Chantel's chocolatier Muhammad Zuhair Odabashi, a man in his fifties, who works for all the firms. During our second visit to the factory, he emerges from the company's kitchen with yellow heart-shaped molds and a bag of strings wrapped in plastic like balls of hashish.

Al-Moutem and the younger Murad brother fiddle with a contraption on an ironing board, then announce that they're ready. The brother holds up a bra and knickers connected by a thin strip of material down the front. Al-Moutem claps his hands, and the outfit falls apart—the bra flies open from the front and the thong drops down in pieces—Ali Baba undressing on command.

Although women sew at Chantel, the brothers design entirely on mannequins. They also rely on al-Moutem. He tells them when a style is too *falahin*—from the word meaning "farmer" or "peasant"—and his suggestions are sewn into the next batch. The models are outspoken as well and veto styles not because they are too raunchy but because they look too cheaply made.

In a globalized market, influences are interwoven. The Syrian firm Farah (Joy) looked to the racy U.S. lingerie firm Shirley to set the standards. In Farah's offices in the modern shopping district of Damascus of Salahiyah, Alaa Kasar, the company's co-owner, pulls out a page from a Shirley catalog he keeps in a desk drawer. On the page are busty blondes in mesh negligees. He came across Shirley at a French trade show and visited the company's U.S. offices. Farah now exports to Holland and Germany, but the company's red and black PVC outfits, with zippers on the crotch or across the breasts, look crude and unimaginative. Kasar admits that his ideas spring from unlikely sources. A meal of chicken and potatoes inspires a black-speckled yellow-feathered bra and matching knickers. A pair of decorative Moroccan slippers becomes a diamanté pattern on a thong.

Good design makes a lasting impression on the souk's buying female public. Amasel Fashion is known for its outlandish styles. It uses another emblem of the Arab street, the butterfly, and dots small cloth ones on bodysuits. Amasel also manufactures "the curtain": patterned material over a wire frame, with a drawstring on the side that opens peekaboo flaps over the breasts and crotch.

The first time we visited the Amasel showroom in Salahiyah's Souk al-Ma'rad, it was a paean to kitsch. Black mannequins were outfitted in Hawaiian wear—paper leis, lei nipple clips, and straw skirts and thongs—and one-piece sheer black body stockings trimmed with red PVC hands grabbing all the right places down the front, each with a matching PVC bra and tasseled panties. Two heavily veiled women were shopping for their lingerie store in the Gulf, and the Amasel salesman quizzed his unexpected visitors in case they were "lingerie spies" looking to steal the company's latest designs.

When we return two years later, the lingerie on the racks is remarkably elegant, including a brown and turquoise paisley bodysuit dotted with gold crystals and a pink printed top and matching skirt decorated with photos of trendy dark-haired girls who look Italian. Amasel's 2005 floor-length chiffon dresses and pantsuits are more suited to private entertaining than getting into bed. The owner, Mouhammed Doukmak, does all the designing, but unlike his rivals, the lanky, red-haired twenty-nine-year-old has had sewing

experience. He used to make party dresses for his mother and sisters when the family lived in Kuwait. After they moved back to Syria, at age seventeen he started Amasel Fashion, whose style of lingerie he feels best articulates the pressures facing Arab women today.

Named after a former girlfriend, the company makes lingerie for women who are in touch with their sexuality. Like many Arab women, the Amasel customer doesn't work and stays at home; her family is important to her. Middle Eastern women, Doukmak explains, live under enormous constraints. "A woman constantly has to keep changing herself in order to entertain her husband." He wants Amasel lingerie to address "what it's like for a woman to change all the time and how she feels when she's trying to seduce—what are her emotions?"

Because of the religious distinction between women's outer- and innerwear, "indoor clothing" has a bigger market in the Middle East than clothes worn in public. Indoor clothing, often bright and garish, counteracts the anonymous drabness of the mainly black, gray, and blue clothes worn in public. As Doukmak observes, "In the West, clothes worn in public are flamboyant. Here flamboyancy is reserved for the home." He takes his Pucci-inspired leggings home to his wife, who wears them around the house while she's cleaning.

Doukmak is the first lingerie proprietor we have seen whose designs make sense. He draws inspiration from history, television, and the Internet. The silver-fringed, fluorescent green go-go set was inspired by the Charleston. The better-quality leaves and flowers that adorn his styles suggest a return to nature and innocence. A snug Amasel pencil skirt is attributed to Nefertiti. Sometimes his obsession comes from an unexpected source. When his son burst a balloon featuring a map of the world, Doukmak stretched the remnants over bra cups and added leaf-trimmed turquoise panties. Beads, he stresses, make any style look better.

On the factory floor upstairs, two spacious, well-equipped rooms are used for cutting and sewing. Under a twelve-foot-long table in the cutting room is a treasure trove of Korean and Italian textiles, including Lycra, the modern miracle material not produced in great quantities in Syria. The only two women in a mainly male workforce sit together, trimming threads from newly sewn thongs. They say Amasel is a good company to work for, as their boss hovers a little nervously nearby. (He's never far away.) Above his office desk downstairs, a closed-circuit monitor surveys the factory floor.

The Secret Life of Syrian Lingerie

Above: In Angel Lady's Damascene basement-factory off Baghdad Street, the fitting of a new lingerie style is done by eye.

Page 22: A woman prepares to sew bikini lingerie from an edible material, which comes in many different flavors including Nescafé (shown) and tutti-frutti.

Page 27: This lingerie shop, Ramadan, named after the owner, sells, in addition to wedding lingerie, cotton underwear for men, women, and children.

Photographs by Reine Mahfouz	Essay by Malu Halasa

Between Souk al-Hamidiyeh and Straight Street—where the great architect of the Christian Church, St. Paul, made his first converts around A.D. 70—lie the choked, tiny streets of Hariqa stuffed with Aleppan wholesale textile outlets, Thai plastic tablecloths, and carpets from everywhere else, among other household wares. Sharing a basement with a *kefiyyah* headdress factory in Hariqa, Lingerie Lour sells sex-themed gift packs featuring a candle, a feather, a small bottle of scented oil, oversize dice, and an instruction booklet. To play, a couple rolls the dice. The resulting number corresponds to an innocent game: a kiss, a tickle, a massage, et cetera. Lour also sells wide-brimmed satin-covered red hats—the same style Muslim brides wear in white—with the words *kiss me* on the front.

Owner Yousef Daeih entertains friends and employees in a basement room crammed with clothes, a small mountain of lingerie boxes, and shelves stuffed with lingerie. One of his friends, Dareen Habloul, is fairly unimpressed with Syrian lingerie. The twenty-six-year-old does not like the jokey styles, although she admits to wearing *ish al-asfour*, the famous bird's nest, but found it uncomfortable. Samir Aboud, a jeweler who designs Lour packaging, says his wife calls the lingerie *spoor,* meaning "cool." A general consensus about the lingerie seems hard to reach.

According to Sumaya al-Ali, who works for Lour and is one of the few female lingerie designers, women have a much more difficult time purchasing lingerie than men. At least men will tell the twenty-eight-year-old woman in a *hijab* what they want. Women, she says, are simply too shy. She takes one of her creations off the rack. It is a two-piece set consisting of a short midriff top and miniskirt made of a material not often associated with bare skin—burlap.

"Women bring something new to lingerie design," says Daeih a few days later in Lour's one-room factory in Jarmana, a new town on the outskirts of Damascus. "But men," he continues, "go into fantasy. They're more imaginative."

In a good year, Lour makes a profit of $20,000—this in a country where the gross national income per capita is $1,190, according to World Bank figures. Most of his customers are from the Gulf. The day before, the company filled a rush order for a hairdresser who sells lingerie out of her salon in Tripoli. Because of yesterday's workload, Lour's main designer, thirty-two-year-old Shadi Sioufi, shuffles into work at 11 A.M. When he states that the inspiration for his designs comes from God, his boss contravenes: "God has no interest in lingerie." He then adds that they look at French magazines like *Cherie* and at U.S. brands like Victoria's Secret.

"European companies are known for their 'classic' styles," says Daeih. He uses the adjective as many lingerie proprietors do, to mean "unadorned." Arabs love ornate, glittery lingerie. He is echoing Doukmak's theory of the flamboyant. At Lingerie Lour, feathers, not beads, inform styles. Daeih holds up a floor-length, midnight blue, sheer housecoat, the neck trimmed with black feathers, and a matching nightgown. It evokes 1950s Hollywood films.

Back in town at Angel Lady, lingerie designer and manufacturer Firas Nabulsi, thirty-six, prepares for Ramadan. A calendar listing the times to break the religious fast hangs in his office. Extra copies are on the cutting table in the factory room next door, where a woman in a white *hijab* pins the distance between two plump heart-shaped red ruffles in an attempt to balance them on a mannequin's breasts. In the West, bra sizing is based on back measurements and the distance from the rib cage. At Angel Lady, dimensions are calculated by eye. Another woman in a *hijab* makes the ruffles, and someone else sews them onto a string. Red chiffon ribbon festoons the machines and the floors. Assisted by a fifteen-year-old boy, the women sewing range in age from nineteen to twenty-nine. The youngest has been working for Angel Lady for only two weeks, the oldest for three years. Do they tell people they sew lingerie? Of course, they say proudly, lingerie is a luxury item. The nineteen-year-old newcomer says she looks forward to wearing lingerie when she marries.

When a tag reading "Made in the SAR [Syrian Arab Republic] Italy Model" is pointed out, Nabulsi admits playing on regional prejudices, as many other lingerie proprietors do. Hence, the profusion of Syrian companies with French names. Other regional preferences are also at work. Angel Lady manufactures outfits that are decidedly of a style that would be worn by a cabaret or belly dancer—a sequined or fringed bra and voluminous low-slung hipster pants with or without additional sheer skirt paneling, topped off with a scarf or veil. These outfits could easily be perceived as Orientalist stereotypes inspired by the lustful dancing girls of Algerian colonial postcards or allure-of-the-East paintings by Ingres. For Nabulsi, however, the impetus behind these outfits is homegrown and religious. "The Quran says a wife should entertain her husband. She should dance for him."

Later, after the interview, Nabulsi and al-Moutem try to argue that lingerie is religiously acceptable because it services husbands and wives, but once they agree this is implausible, they become circumspect. Despite his work, al-Moutem says he doesn't believe in premarital sex. For Nabulsi, the problem

for him as a Muslim is extramarital sex. In his twelve years of selling lingerie before setting up his own company, he learned that men are freer with their mistresses than they are with their wives. They buy elegant loungewear for their wives, but the more experimental styles go to their mistresses—something not unique to the Arab world.

Still, the question remains: How sexually arousing can a thong with a Tweety Bird playing a loop from Kajagoogoo be? Nabulsi is pragmatic. Because his customers rarely see a naked woman, anything she wears is exciting. Lingerie has its practical uses for his female customers as well. He concludes that it keeps husbands away from prostitutes.

Prostitution seems to be a recurring theme, and many of the Syrians we talk to, both men and women, bring it up. With the ongoing war in Iraq and the continuing unrest in Beirut, Damascus has become a sex capital, so much so that Iranian delegates securing a special relationship with Syria's current president, Bashar Assad, complained about the fleshpot nightclubs and nearly naked women performers during Ramadan television spectaculars.

3

The relatively few uncovered women on the streets of Syria's second-largest city, Aleppo, give the impression that the headscarf is compulsory. Yet, consistent with the distinction between public and private, interior spaces say more. Aleppo is also filled with female gyms and sports halls where women of all religious persuasions mix freely. These are the new *hammams*. Palig Avakian, a thirty-five-year-old Armenian Syrian and a mobile-phone customer-service representative, trains three times a week and boxes. "As a woman, I can enjoy my body, and veiled women can enjoy their bodies, too," she says.

In the labyrinth of Aleppo's Old City, lingerie shop owner Nihad Tabbakh dates the buying and collecting of racy lingerie to the Tishrin or October War. In 1973, Syria and Egypt launched a surprise attack against Israel during the Jewish holiday of Yom Kippur. Although the Israelis pushed back the Arab armies, the war is considered a success in light of the 1967 Six-Day War, when Syria lost the Golan Heights. After 1973, the Gulf countries started investing militarily and financially in Assad's regime,[1] and luxuries like lingerie began circulating.

Photographs by Reine Mahfouz   Essay by Malu Halasa

Tabbakh remembers: "We were at war, and you cannot celebrate when you're afraid. After a war, there were many weddings, and women began to wear sexy lingerie."

A doctor friend of Tabbakh's visiting the shop suddenly excuses himself, saying he is in a hurry to go to the mosque for midday prayers. Within minutes, Tabbakh's coworker opens a Quran on the counter and begins reciting under his breath. Tabbakh, in his sixties, made the hajj. For someone so religious, he describes the selling and buying of lingerie as uncomfortable for all concerned: "The young ones are rather embarrassed. According to Islamic ideals, a woman must stay at home and cannot associate with men, but she wants to show her husband that she has something special. Sometimes they mention that they are looking for thirty colors in thirty different outfits. They want many colors because they don't yet know which color their husbands like."

The history of lingerie is part of the alternative economy of the Middle East. At the same time Syria was buying weapons from the USSR in the 1960s, cotton manufacturers like Shenineh in Aleppo were shipping tons of rudimentary and ill-fitting cotton underwear, T-shirts, vests, and briefs—but no bras—to Russian cities.

The bra, a feat of Western fashion engineering, developed in the social evolution of more active lifestyles for women. Due to a variety of reasons—border closures, corruption, and scarcity of modern materials and machinery—the bra was not manufactured in Syria until the 1970s. Prior to that, if women couldn't afford to purchase foreign-made bras, they resorted to other measures. Shenineh's oldest employee, sixty-four-year-old Abu Ali Salah al Din Belal, is like many Syrian husbands; for the past forty years, he has done all the shopping, including the purchase of underwear for his wife, who stays at home. He remembers the time before the bra, when women used to strap up their breasts with thick elasticized bands, much like the headbands used under the *hijab* today. Another alternative was wearing a full-length slip, called a *shel-ha*, with an embroidered top and a skirt finished with an elastic hem. From 1940 to 1960, the garment stayed the same. In later years, it was shortened to keep pace with changing hemlines. Now Abu Ali's wife asks him to bring home peach- or rose-colored bras.

By the 1980s, European brands realized Syria was a source of good, cheap cotton, and Shenineh started manufacturing more modern cotton styles for French, German, and Italian firms. This was at the time when Hafez Assad lifted restrictions on the country, and the Syrian domestic market, exposed to foreign influences, started clamoring for more variety in underwear styles.

One of Shenineh's owners, Yasser Shenineh, remembers, "Women no longer wanted big panties that came to the waist. By 1987 they weren't asking for the 'string' [thong] but they were looking for something pretty." His factory uses cotton yarn to produce four thousand pieces a day—women's camisoles, thongs, and G-strings, plus men's briefs and T-shirts—destined for the regional and international markets. Cotton lingerie forms part of the country's transition from Pan-Arab socialism to an economic system dominated by the commercial monopolies of the ruling elite—though the modest economic reforms made by the current president appear to give a little more room for the entrepreneur to maneuver.

Yasser continues, "We know our product and our product is good. Where we fail is in marketing." Sometimes the company hires Lebanese models to mimic Calvin Klein poses for brochures, but it would never resort to Eastern European models because, as he believes, it would send the wrong message: cotton lingerie is more about utility, not outright sex. Still, pictures of semi-clad Eastern European women grace the windows of lingerie shops in Aleppo. Since the models are foreign, there is an understanding that their lack of attire doesn't matter because they're not Muslim. According to al-Moutem, Syrian women never model racy lingerie. In Aleppo's cotton lingerie companies, a few women model infrequently, going against family and religion to do so.

Issa Touma is a photographer and curator of Le Pont, Aleppo's only gallery for contemporary photography. The gallery is known for showing "all things scandalous—pieces by Jewish artists, portraits of nude men and women, videotaped performance art verging on the pornographic—none of it submitted to government censors for approval."[2] From 1996 to 2001, Touma earned a significant portion of Le Pont's operating budget through advertising photography for the city's cotton lingerie firms. He has shot ad campaigns for cotton lingerie using Syrian models, but not without problems. When he wanted to photograph outdoors, a company owner became nervous. Touma explains, "Usually on a shoot, the company sends two women to style the lingerie. When we shot outside, the only person who came with us was the driver. I'm sure he was told, 'If there are any problems—run away.'"

Photographs by Reine Mahfouz　　　　　Essay by Malu Halasa

In souks across Syria, women buy their intimate apparel from men.

4

There is an extremeness about a design culture that shows women's genitalia embellished, cradled, encrusted, feathered, or made more delicious with candy and chocolate. Syria's racy lingerie embodies both fantasy and frustration.[3] Veiling does not neuter women, but curtailing the visibility of the body in public often strengthens the desire for it in private.

That is the public side of the story. Inner spaces are much more tantalizing. It takes a certain amount of bravado and chutzpah to slip into a Charleston-inspired silver-fringed bra, thong, and leg- and armbands, articles of clothing that reveal much about the changing sexual mores of the modern Middle East.

Opposite: Kitschy and inventive, a display sells thongs in Lingerie Baqdounis.

Photographs by Issa Touma
Portraits of Women

Issa Touma, a self-taught Armenian photographer and contemporary photography curator, is a controversial figure in Syria. In his article "Artists Still Have Hopes," on the Web site SyriaComment.com, he writes," Unfortunately few artists are willing to take the risks involved in producing images that are explicitly linked to the day-to-day realities of life in the region including political and social oppression. At this point in time, *hijab*-covered ladies' conditions are far too intimidating for significant numbers of artists to face the consequences."

Sometimes mistaken for a tourist with a camera, he is often invited to the roofs of houses to photograph young women who want pictures of themselves without their head coverings. His willingness to dare and not bow to the authorities has meant that Touma continues to have run-ins with Syrian Mukhabarat, or security authorities. Although there is a sense that some progress is being made, change in Syria comes in fits and starts.

Essay by Noura Kvorkian
On Love and Lingerie

In 2001, filmmaker Noura Kevorkian went to Damascus to make a documentary about Syrian lingerie and the women who wear it. She became friendly with a group of women living behind Souk al-Hamidiyeh in Damascus. Originally born in Aleppo, Kevorkian—who is Armenian-Lebanese—grew up in Lebanon. During the Lebanese civil war, her father was forced to close his factory and relocated the family to Syria. In her teens, Kevorkian moved to Canada. The film *Veils Uncovered* was a homecoming of sorts. In a journal kept during the filming, she writes of Syrian women speaking candidly about their lives.

29 July

My first day back in Damascus. I pack my camera and walk toward Souk al-Hamidiyeh. The city is vibrant and the souk electrifying with the hustle and bustle of people. Its distinctive aroma comes back to me—of ancient caravans, spices, rose, jasmine, *ghar* soap, and *mulabbas* candy. Everything feels familiar, as if nothing has changed since the spring of 1978 when I first walked, as a child, through this market, holding my father's hand.

Covered with black cloth and in some parts with perforated tin metal sheets, the souk has many corridors and mysterious alleyways leading to sections of the market specializing in food and textiles. There are antique and jewelry stores, woodcraft and brass-carving artisan shops, and carpet weavers. The sidewalks are lined with *bastars*, blankets, and tarps spread on the ground displaying everything from watches to scarves and musical instruments. Wooden carts display colorful toys, multicolored and beaded bras, and undergarments. Sometimes donkeys pass by, and men's voices can be heard yelling out the names of spices. Children stand in front of candy stalls, mesmerized by the colors and textures of delight.

Standing out is the paradox of fully veiled and burka-clad women shopping publicly for sexy, colorful, playful, naughty lingerie, displayed throughout the souk in windows and stalls and on sidewalks.[1]

Why would a veiled woman buy crotchless panties, underwear with zippers and feathers, or panties with cartoon birds and musical buttons? Is it possible for a woman to walk comfortably wearing underwear with feathered Tweety Birds the size of apples resting against her crotch? Most likely these indulgences are meant to be worn only in the bedroom. But what about the musical panties playing Western pop tunes? Why would a society that shelters its women design and sell such explicit lingerie? Why do the women buy it?

I grew up Christian in the Middle East. These women were my neighbors. Yet a number of contradictions about their lives perplex me, and I am determined to find out their meaning.

30 July

Since breakfast, I've been walking around the souk with my video and still cameras, pretending to be a tourist, absorbing the mood and the life of the market, conspicuously filming the women buying lingerie. I snap photos and approach women I think might be courageous enough to talk to me. It's very slow going.

7 August

Breakthrough! Umm Fathi invites me to her house. I tell her I am interested in women's stories. We sit in the women's room on the carpeted floor decorated with cushions. Her daughters Safai, thirty-one, and Fatima, eighteen, come to greet me. Khadija, her daughter-in-law, prepares coffee.

Umm Fathi is charming, small-built with big brown eyes and a warm smile. At forty-five, she has nine children and fifteen grandchildren. She was married

Photographs by Issa Touma Essay by Noura Kevorkian

Above: After breaking the evening fast during Ramadan, people flood into Aleppo's bustling old town district of Tellal.

Page 34: In the Kallase neighborhood of Aleppo, a woman walks past posters of Syria's dictator Hafez Assad (left), and his two sons, one of whom is the current president, Bashar Assad.

The Secret Life of Syrian Lingerie

Occupants step into the courtyard of a traditional Arabic home that is walled and windowless on the outside.

at fourteen to her first cousin Abdo, who died a few years ago. Umm Fathi is respected as the head of the household. Her grandchildren kiss her hand when they greet her. Her two eldest sons, Fathi and Ali, share the modest family home along with their own families. The house is located a few minutes' walk from the souk. It boasts a small but lovely front patio with blue tiles and a small pond surrounded with flowers planted in rusty cans. They are poor, yet very generous and hospitable, like most Syrians I have known.

Umm Fathi is surprisingly direct: "Safai and Fatima are living with us now. They were sent home by their husbands. Safai is divorced, but Fatima has a two-year-old son and we're trying to reconcile them."

Khadija brings in perfect Arabic coffee. I finish my last sip and look at Safai for permission to turn my cup upside down. Safai nods: "Umm Fathi is a great fortune teller."

Umm Fathi looks into my cup and says I have a long journey ahead of me and that after the second moon I will give happy news to my parents: she sees an *aris*—a groom. As she shows me what looks merely like a lump of black coffee grounds stuck to the side of the cup, the women laugh, giggle, utter numerous rounds of *"insh'allah,"* and ask me to remember them at my wedding. The rest of that afternoon, we talk about marriage, husbands, and divorce. In fact, all conversation seems eventually to turn to marriage.

Umm Fathi's sister Muna comes to visit. After introductions, she sits across from me and removes her veil, making herself comfortable. She is a lovely, talkative mother of seven children, and she is only thirty-two. She says, "Men and women do not interact, so girls and boys look forward to getting married. Premarital touching, kissing, and other *haram* things like that are not allowed, so everyone is preoccupied with marriage and sex." She laughs nervously but triumphantly, as if she's said something that's normally forbidden.

They are curious about why I am in Damascus. What is so interesting about their lives that I have come all this way to talk to them? They want to know how a "free" woman feels, traveling without restriction, without a husband or male chaperone. I answer endless questions about Canadian women, men, dating, about going without a veil. Umm Fathi does not approve of allowing women to live alone. She calls it "Western corruption" and concludes the conversation with: "*Insh'allah*, a nice man will marry you. You'll give him many boys. *Insh'allah! Min timmi ila abua' al-samah* [From my mouth to the gates of Heaven]."

I take my leave and promise to be back in a few days' time. Umm Fathi kisses me three times, saying, "We're always home, we hardly go out, just knock on the door, it's always open for you." Safai waves good-bye.

12 August

At the souk I meet a group of peasant women from a village outside Damascus shopping for the young bride accompanying them. They are very good hagglers. They allow me to film them as they negotiate lingerie prices with Muhammad, a twenty-two-year-old blond salesman.

Back at the hostel, I look out the small window, thinking of Umm Fathi and her daughters and daughters-in-law. They are all veiled and wear burkas whenever they leave the house or are in the company of men other than their fathers, husbands, and sons. Umm Fathi has worn the veil since her wedding day. She says she is used to it and likes wearing it: "I'm a good Muslim. Besides, it makes me feel protected."

But Safai, a real beauty with a slim athletic body, a lovely smile, and sparkling eyes, does not seem so convinced: "God gave me a pretty face, why can't I show it? Why do I have to hide it from people? Why can't I go out whenever I want?" Safai knows the Quran does not require women to cover up, but she does not have the power to fight for what she wants. As a divorced woman with no financial support, she relies on her brothers' care and must obey their wishes. Brought up in a more liberal family or belonging to a different economic class, Safai would be an outspoken woman, a fighter and a leader, exploring life to the fullest.

Safai refuses to talk to me on camera. She is scared about her reputation, of dishonoring her brothers. She needs them.

13 August

Since this morning I've been filming in the souk, recording scenes of veiled women buying lingerie. I try to film a young woman accompanied by three older women. One of the women blocks my way, refusing to allow me access.

In the afternoon, I enter a carpet store. Two middle-aged men sit on small wooden stools playing *tawleh* (backgammon). Another two are sitting on a carpet, restoring kilims with colorful threads and large needles. They are friendly and offer me coffee. I accept; it is a chance to talk to them about the women who frequent the souk.

However, I'm unable to get a definitive answer from them about their thoughts on lingerie and why veiled women might be inclined to purchase racy undergarments. Do their wives buy such lingerie? The men seem to know exactly what to tell me without saying much. They are very polite and respectful of women, and I get the feeling that at least one of them has a veiled wife who buys sexy lingerie, or a lover who wears lingerie for him. Still, they tell me that relations between husband and wife are sacred, not to be discussed with other people.

They also explain that sometimes men can get distracted by bad women who possess them and lure them with clothes and dance—the work of the *sheytan* (devil)—but they always go back to their wives and children. They try their best to say what they think I want to hear. They forget I'm a Middle Eastern woman, that I grew up in an Arab culture where men customarily lie, commit adultery, and get away with almost anything. I remember my grandmother telling me: "A man who grazes on streets eventually returns home to his wife. A wife must be patient and close an eye on things she does not want to see. It's better not to see or hear. What else can she do? Once a woman has kids, she's tied down and has to swallow even shit to keep on living." My grandmother had a way with words.

14 August

I don't feel like getting out of bed. I feel frustrated. I've been here over two weeks. Although I have met and chatted with many people, I still don't have an actual interview on tape.

Today is the day. I have given them plenty of time to get used to me. I will ask my questions today.

Fatima opens the door and welcomes me with a smile. We sit in the *aghda* (sitting room). Safai and Muna come in from the kitchen, and shortly afterward, Umm Fathi returns from the neighbor's house. As we're drinking coffee, I ask Umm Fathi if she ever buys lingerie from the souk. Surprised and a bit nervous, she laughs, saying, "I don't have a husband anymore."

I tell them I have seen many varieties of strange and exotic underwear in the market, and wonder who buys them. Muna says, "Only married women buy lingerie for their husbands. Women don't buy it for themselves—who wants to wear uncomfortable plastic panties? But we do what we have to do." She continues, "A woman's job is to please her man. He has to be happy and satisfied at home, otherwise he would go out . . ."

The Secret Life of Syrian Lingerie

With a panorama of Aleppo behind, a family steals time on the roof after a long winter.

Photographs by Issa Touma   Essay by Noura Kevorkian

In the privacy of her home, a young woman and her female relatives (not pictured) forego the head coverings they wear in public.

I play stupid and ask, "Where would a nice Muslim man go in Damascus?"

"*Haram*, poor thing, you are so innocent! If you were married, you would know that if you don't keep your husband happy in the bedroom, he would go out—whores, mistresses, and worst of all, he could marry a second wife," concludes Muna.

The two divorced sisters react differently. Safai laughs nervously while the sad Fatima becomes lost in thought. When I ask Safai what's so funny, no one answers me.

16 August
Back in Beirut, I sit in an Internet café verifying what I know about Islam from my university studies and looking for recent statistics. Polygamy is permissible. The Quran is clear: A man is allowed to marry up to four wives at one time, but must love, respect, and provide for them equally. No favoritism.

But the Quranic laws are made more complex due to the heterogeneity of Muslim societies and cultures. Even within a homogeneous culture, they would prove complex due to differing social and economic classes. In the Syrian middle and upper classes, women are educated and professionally employed and, to some extent, participate in political life. In these classes, polygamy is almost nonexistent. In the uneducated lower and rural classes, polygamy does exist, however uncommon it may be.

Regardless of the actual number of polygamous marriages and the social status and education of the men and the women involved, it becomes clear that polygamy is used as a threat: "If you are not a good wife, if you don't do this or that, if you don't give me sons, I'll marry again."

It seems obvious that the sole purpose of lingerie is to seduce partners, whether one lives in the East or West. So it's not news to me that the lingerie in the souk is used to stop husbands from straying, but I did not expect to find so obvious a connection between lingerie and polygamy.

I take a taxi to West Beirut to visit my friend Janine at the American University of Beirut. I tell her about the women and the stories I heard in Damascus. She smiles and takes me across the palm-tree-lined boulevard to a bookstore. Janine talks to the bookstore owner, and a moment later he hands me a book. That night, I read this poem:

The Second Wife
The stranger has come; she has her place in the house.
Her tattoos are not like ours,
But she's young, she's beautiful, just what my husband wanted;
The nights aren't long enough for their play,
But we'll soon see if she's as good for work.
Now she's dolled up in silks and scarves
In colors brighter than *bagzoua* feathers,
But soon her skin will have to get used to rough wool,
To carry firewood, milk the cows, and cook.
People are criticizing my husband,
Not for taking a second wife,
But for bringing a stranger to our land,
A stranger we don't know much about.
Some say her parents were of Ait Tarkebout;
She can't be a daughter of the Ait Tarkebout.
She claims she's of Ait Bou Ou Guemmez;
I think she's probably from Ait Bou Demmez.
Since she's come, the house is not the same,
As though the doorsills and the walls were sulking;
Perhaps I'm the only one who notices it,
Like a mule before his empty manger.
But I must accept my new lot,
For my husband is happy with his new wife.
Once I, too, was beautiful, but my time is past.[2]

22 August

Back in Damascus after spending five days with my family in Lebanon. From the Garage Lubnan, I hop into a taxi and go visit Muna, Umm Fathi's sister, who is expecting me. Safai is meeting us there. Muna has called one of her friends, Huda, to meet me as well and also arranged for me to visit her neighbors, Leila and Soheil, who have agreed to speak on camera.

We sit in her kitchen. The men are at work. Her mother-in-law is sick and sleeping. The children are in school. The radio plays a dance song with a fast tabla beat. The coffeepot is on the stove. Huda comes in and shakes my hand.

As she removes her headscarf, she looks at me playfully and sings: *"Noura, Noura, ya Noura, ana bahibik ya Noura* [Noura, Noura, oh Noura, I love you, oh Noura]."[3] We all laugh.

Huda is a large, beautiful woman. She weighs more than two hundred pounds and has radiant cheeks framing dark eyes and big, fat hands. She is talkative and funny and loves to dance. I pull out my camera, but Huda refuses to be filmed. Too bad. She is so photogenic and captivating.

Huda is married to Ala', her seventh husband. Her first three husbands died, and two others divorced her. Unusually, Huda took it upon herself to divorce her sixth husband. "My current husband, Ala', may God keep him, is slim and short, half my size. He is a nice man, but very shy."

Huda's friends jokingly tell me she is the boss in her house. I ask if that's why she doesn't wear a veil.

"When we married, I told him I did not want to wear the veil. And he agreed."

"Because he was afraid you would kill him, too," adds Muna, mischievously.

All of us are in a good mood. Stories flow. I ask Huda to tell us about lingerie with respect to her seven weddings and seven husbands.

"One week before my wedding," Huda begins, "my female relatives took me shopping in the bridal area of the souk. There we bought my bridal gown, a few nice dresses, and a lot of lingerie—different sets for different occasions: white sets for the wedding, red ones for the *dekhleh*, the first night of the wedding, and colorful sets for the honeymoon. Usually a bride's lingerie collection includes lovely silk and satin long nightgowns with matching robes, pretty shawls and panties, colorful bras, and at least one belly-dance outfit. Every woman needs to dance for her husband."

When I ask Huda if she dances for Ala', she laughs nervously, like a shy girl of fifteen. She shakes her head. I tell her I love belly dancing, but that my father, like the rest of the Middle Eastern men I knew, did not allow me to dance publicly at weddings or parties. It was considered taboo, a dance for "dirty women."[4]

"Men say it is dirty, that it's for the whores," offers Muna, "but then they ask us to be their own private dancers." How I wish I had these conversations on tape.

"*Eh wa-Allah!* [God knows that's right!]" says Huda.

Photographs by Issa Touma   Essay by Noura Kevorkian

Above: On New Year's Eve, a belly dancer entertains a Christian wedding in Sissi House in Aleppo.
Pages 46–47: Women stroll the wet streets of Raqqah, the last main town near the Syrian-Iraqi frontier.

The Secret Life of Syrian Lingerie

In the surrounding villages, agricultural laborers travel by tractor.

Safai interrupts, "Like *jozi al-himar* [my donkey husband], no matter how much I took care of him, how I pleased him, no matter how many dances I danced for him, he married that *sharmuta* [whore]!" Muna reprimands her niece with a look that says *sharmuta* is not a nice word to use.

Safai's marriage to her second cousin was arranged when she was fifteen. She calls him *jozi* (my husband) or Abu Najem (Father of Najem), Najem being her eldest son. Gradually she began to love him, and in time they established a good marriage and had children. She was obedient and did everything the way he wanted. "He treated me well, bought me a ring and a gold bracelet when I gave him a son. He loved nice undergarments and gave me money to buy lingerie."

Safai was the first woman I met in Damascus who admitted to a sex life she actually enjoyed. Her husband's business went well, and they moved out of his parents' house into a small apartment across the street. They became more intimate, spending time together, and she gradually became a woman.

"I love Umm Kulthum, Abdel-Halim Hafez, and Egyptian Ra's music. After dinner Abu Najem would sit on the bed with a cup of tea and sometimes *arak*, and I would dance for him wearing my belly-dance outfit. We had a lot of fun. He was harsh when I showed too much desire or did something new. He would slap me, saying, 'Where did you learn that, you whore?' So I learned to be a good wife and waited for him to tell me what to do. If it was his idea, then we were okay. Men are like children, I'm telling you, wait and see. *Insh'allah* you'll get married one day, then you'll say, 'Yes, Safai was right.'. . .

"For three years I knew he was having an affair with that woman. And then he married her! I hate her. She is evil and ugly, *mitel al-gird* [like a monkey]."

Safai could not stand the second wife. They were always fighting. She felt hurt and jealous. Finally, two months into his second marriage, her husband declared her divorced and sent Safai packing to her parents' house.

When I ask Safai how she was able to stand living with him for three years knowing he was having an affair, she answers, "What could I do? I have children. I needed him." She turns away and starts crying.

Huda jumps in, "It is better to have a husband than be back at your parents' house, at the mercy of brothers and their wives."

"This is the way our life is here," adds Muna.

Huda tries to change the mood. She turns on the radio, positions herself in the middle of the room, and starts dancing. Her big fat hands clapping loudly, she laughs and calls me to join her.

That afternoon we go to the neighbors' house and meet the women Muna has arranged for me to interview. I finally manage to get my first interview with a veiled woman on tape! When Leila removes her veil, I fall in love with her.

Back at the hostel, I think of Huda and of the interview I never taped, a conversation about how she divorced her sixth husband, Ali. I must write about this before I go to bed.

Ali married a second wife and had been spending a lot of time with her. He had not visited Huda's bedroom for over three months—reason enough for a woman to ask for a divorce lawfully, but with one catch: she would not get the *muakhar*, the postwedding contract payment.

While in the Christian Middle East a divorced woman is undesirable for remarriage and is vulgarly labeled "used material," in Islam, a divorced woman almost always manages to remarry shortly thereafter and with no taboos attached. Is Islam more sympathetic and humane in its treatment of divorced women? Can a culture so oppressive to women be modern in its approach to their sexual needs? The Quran stipulates that it is a sin for women of childbearing age to remain unmarried. Which is the underlying basis for such values: the woman's sexual rights or her duty to bear children?

11 September

At Umm Fathi's house over coffee, we're talking about the plan for tomorrow, the big interview day. Muna, Huda, Leila, Soheil, and maybe other women will arrive at the house. With Safai, Fatima, and Khadija—even though the last two are shy and hardly speak—we'll be a group of eight to ten women. I will film the gathering. Some women have agreed to be filmed on the condition that they keep their veils on.

We have lunch, a wonderful eggplant dish with rice. After tea I rest on the divan and listen to the radio, waiting for the men in the house to leave. It seems best to come back later. I say good-bye and walk around the market.

Sometime that afternoon the chaos begins.

I hear that airplanes have crashed into the Twin Towers in New York. Every radio station in the souk airs the news. The atmosphere is confusing. Will the U.S. bomb the Middle East?

I take the first taxi I find to Garage Lubnan. I leave Damascus in a rush, worried about the political situation. Having once lived through a war, I expect the worst. I'm scared and want to be with my family.

I want to call Umm Fathi and the women, but they don't have a phone.

In the taxi, I start to panic. What if war breaks out? What if the men in the souk turn out to be right, that the U.S. has already sent jets to bomb Damascus? I think of the women and fear I may not see them again.

As the taxi speeds along the Tarik Sham Highway, I miss Safai and Huda already. I miss being surrounded by women so innocent yet so wise, so loving yet so unloved, so generous yet poor. I feel confused. Sometimes I envy them and want to be like them. I want to live with them, be part of their sisterhood, their support network. It all seems so uncomplicated, so loving and warm. Other times I feel sad for them. I want to give them their freedom, offer them power and knowledge. Yet I cannot help them. Even if I could, would I have the right to change the way people live, their culture, their traditions?

I feel lost, desperate to figure out these feelings. But one feeling I take away from my time with these women is respect. I admire their courage for being able to get up in the morning, smile, cook, clean, go about their obligations in life—responsibilities to their families, their religion, and, most notably, the challenging obligations they have to their husbands. I admire them for being able to put up with their husbands' demands, wear uncomfortable and sometimes degrading lingerie, fight for husbands they don't even love.

As the taxi speeds toward the Syria-Lebanon border, I feel lucky to have come to know these women who are hidden from the world, to hear from women who are otherwise unheard from.

I plan to return to Damascus the following week, once the political situation settles down, to continue recording their story.

I never get the chance.

August 2004 Postscript

I returned to Canada in November 2001, having decided to make the film based on the footage I had. Although *Veils Uncovered* is not the film I had originally envisioned, it has certainly facilitated discussion!

Returning to the Middle East for the first time since 2001, I review the last few years following the release of my short documentary. Although well received, the film angered Muslim individuals and Syrian political groups, who claimed it stereotypes Muslim women. Yet I have taken pains to clarify that my film documents a small group of women living around the souk. They are relatively uneducated, lower-economic-class citizens who, by and large, support a successful lingerie manufacturing industry. However, it is important to note that not only is lingerie designed, made, and sold in Souk al-Hamidiyeh, but it is supplied to every corner of Syria, from the cities to the rural areas. So is lingerie a common denominator that speaks about the way of life of Syrian women?

Stepping out of the taxi, I find myself back in Damascus, standing in front of the souk. But this time I have my new *jozi* in tow—Umm Fathi was right after all! I have not brought my camera. I will try to contact my friends again: Muna, Umm Fathi, Safai, Huda. Someday I may finally get the interviews. Now, however, I'd rather just talk to them about what it means to be married, and what lingerie they might advise me to buy.

Photographs by Issa Touma Essay by Noura Kevorkian

Only moments before on the back of the motorcycle, a woman makes last-minute arrangements in a busy Aleppan street.

Interview with Ammar Abdulhamid

A Fundamentalist Changes His Mind: Sexuality and Humor in Syria

Political commentator and novelist Ammar Abdulhamid is considered one of Syria's "daring modernizers."[1] The protagonist of his first novel, *Menstruation*—a young Muslim fundamentalist who can smell when women are having their periods—was unexpected within the context of the Middle East.[2] Abdulhamid, son of movie director Mohammed Shaheen and actress Mona Wasef, grew up in Damascus and attended the University of Wisconsin. Returning to Syria in 1994, he founded DarEmar, a publishing initiative devoted to raising Arab civic awareness. He also runs DarEmar's Internet research forum, al-Tharwa (Arabic for "wealth"). An avid blogger, he reports on his country's political upheavals on "Amarji—A Heretic's Blog."[3] In 2005 he returned to the United States as a visiting fellow at the Saban Center for Middle East Policy at the Brookings Institution.

MALU HALASA: *The main character in* Menstruation *is a fundamentalist in his twenties who has a twisted relationship with women. Where did the idea come from?*
AMMAR ABDULHAMID: I was going through a fundamentalist phase in the United States, attending classes at a mosque that dealt with different issues. There were only men involved, of course. One day, the issue was how women should deal with their periods. Although most of the men in the class were bachelors, sooner or later we were going to get married, and we needed to tell our wives these things—as if their mothers wouldn't have told them!

Menstruating women are impure; they cannot touch the Quran, pray, or fast. It is a stigma. In olden times, the teacher told us, Islamic scholars took [the then-equivalent of] a tampon to the imams, who would inspect and smell it to determine whether a woman was pure or not. This was 1,400 years ago, when Islamic jurisprudence was being worked out.

Even then, as a fundamentalist, the whole idea sounded macabre. The idea of classifying women according to fluids was ridiculous.

So contrary to Western expectations, in Islam women's bodies—what they're capable of doing, and their fluids—are well outlined, discussed, and specified. Women are not mysterious.

At the same time, why is menstruation considered impure? Isn't it simple biological reality?

The more experience I gained in life, the less fundamentalist I became. My fundamentalism was a reaction to a fear of life, of reality. The more confidence I gained, the more I questioned everything.

I took a class in anthropology at school, and we were taught about tribal cultures in Africa and Australia, and also about the Babylonians and Assyrians, where menstruation was not considered a stigma. It was a sign of life in agricultural societies, where the deities were female and women took a more active part in religion and the decision-making process. They celebrated women when they reached maturity. The value judgment was positive: women menstruated because they could give life.

At the same time, in some tribal cultures, men regarded menstruating women with jealousy because women could give life. They saw that women started bleeding, which meant they could bear children.

The magical power of menstrual blood was also something they believed in. In some Australian Aboriginal tribes, when boys reach a certain age, in order to mimic women's periods they take a knife and split the penis, letting the boy bleed as a sign he is menstruating too—he is a man now. Fascination with the phenomenon is very old.

Looking at these matters was an eye-opener for me. Many questions in my mind about Islam, religiosity, and sexuality were answered when I studied anthropology.

When you describe yourself as a "fundamentalist," what do you mean?
I believed Islam is right and all other religions are wrong. I wanted an Islamic state. I believed in the conspiracy against Islam. I started dabbling with fundamentalism at nineteen or twenty, but the situation got worse when I came to the United States and this culture shock took place. Religion empowered me. Instead of saying "I'm shy," I could always say "I'm religious" vis-à-vis women.

You had the same reaction to the West that Sayyid Qutb had—except his fundamentalism became the philosophical basis for a violent religious movement.[4] Experience of the West seems to affect young Muslim men badly.

I've had that love-hate relationship with the West since my teens, when I went to England on my own. Western culture is so attractive, but so different from anything we're taught [in Syria]. The trouble with our culture is that even liberals have an issue with sexuality. I was not taught that sexuality was okay, even in my liberal family background. My parents didn't have time for that. I learned on my own, picking up things by myself.

Is that the case for young Syrians, as there obviously isn't sex education in schools?
You could say the culture teaches us everything. Whether a woman wears a veil or not, or whether men and women are segregated or not, there is always contact between men and women, always this need for contact. Nobody marries at an early age anymore, yet we are all expected to remain virgins until we get married. Many of us do not marry until our thirties or midthirties, so the expectation is unrealistic. Different levels of sexuality exist in Syria, but there is guilt involved. People try to sideline the issue, or say "God will forgive." They are not reconciled to their sexuality even when they break Islamic codes and indulge in sexual activities. This creates a lot of negative reactions later on, when a man marries, how he treats his daughters, or how a boy treats his sisters—because he knows how he thinks about women.

Arab women seem to be objectified on many levels, East and West. There is a perception of the way women ought to be in Islam. There's also a perception in the West about the way Muslim women are because of how they dress, or because of their religion. Then there's the reality of what women are permitted to do in their respective countries, which can vary greatly due to economic, social, and political conditions. That's what makes Syrian lingerie such an interesting light on sexuality. We do not expect it to be there, so open, so available, but it is.
My interpretation is that women's ability to give life is a source of envy for men. So there was the inevitable development: how are we going to control this fountain of life? You classify them. When you classify something, you have power over it. I think this is the sole idea behind classifying women into pure and impure, also imposing the veil or segregation. It's all about control. If you control women's bodies, you are halfway to controlling their spirits.

The stories of *The 1001 Nights* tell a lot about how women behind closed doors conspire to get power back. In *Menstruation* the bedroom scenes are about how women try to regain a little control, by having these kinds of relations and exploring their sexuality.

What is amazing about the country's lingerie culture is its sheer variety. There are a multitude of very racy, very sexy styles, and it takes a certain kind of confidence for any woman, regardless of religion, to wear them.
It's true. On one hand you're turning women into sex toys. They're not supposed to be sexually stimulating to other people, but at home, to the husband, they're supposed to provoke his sexuality and dress in the manner that will attract him, and do whatever he says. At the same time it gives women a lot of control. Women can use sexuality to manipulate men.

Are there constraints on women in public life in Syria?
Not as much as in Saudi Arabia. In Syria it is more complex. Women can drive and work. Whether they wear a veil or not does not stop them from getting an education. The issue becomes: this veiled woman goes out and works, but at home she still has to follow the man's decisions. Being a contributing economic player at home does not mean you're treated with equality. On the other hand, you can be a woman, you can wear a bikini and a miniskirt and drink alcohol and have a freer sexual life.

But at the same time the culture teaches you that a man should knock on your door and ask for your hand in marriage. This was the shocker for me when I went back to Syria. Many women who might not wear a veil or who might even be sexually active still expect [that] knock on the door, and then haggle about the dowry—what are you going to give?

When you say "knock on the door," are you referring to an arranged marriage?
Marriage is not necessarily arranged by the parents, but you have to get parental permission. Parents can veto a marriage even though the girl might be in love. The material qualifications are still paramount. What the boy offers is important—the car, the house, et cetera.

If these women are not main economic players in their own lives, isn't marriage about securing a woman's future?
Not necessarily, because if you look at the divorce rate, it is more than 50 percent in Syria. There are no official figures, but there is evidence from lawyers.

So Syria is a traditional society in rapid transition.
The problem is that, in our culture, modernity did not come piecemeal. Western societies had a long time to think about issues of gender equality, women voting

and working, sexuality, birth control, homosexual rights, elderly rights. These ideas were already developed, and Western societies absorbed them gradually. But in our societies all the ideas were put together in one package, and through TV and satellite TV, very traditional people are exposed to postmodern culture. They can see all these issues on sexuality, women's freedom, women in control and out of control. To them it's shocking. This has complicated things. In our culture you meet people who are very liberal, Westernized figures in their way of life or convictions, and you meet very traditional people. You meet veiled women who are more staunchly feminist than in the West. You also meet veiled women who are very traditional.

It's hard to make generalizations about a culture that is so complex. Yet the United States makes massive generalizations about Arab societies and has acted on them to devastating effect.
This is exactly their mistake. All cultures are in flux, but this is a traditional culture facing a postmodern culture. All the fermentation that took place in the West in the span of a few hundred years is taking place in a span of decades. If you take a cross section of Syrian culture, you are going to see a spectrum of different cultural values that cross a thousand years, but they all exist right now in one single moment.

From my travels in Syria over the last five years I can tell there have been enormous social changes. There are pubs, satellite TV, music videos. The way young Syrian women dress has changed in the last two years. You see young women in tight jeans and skimpy shirts, but also in hijabs.
You can wear tight jeans and a tight T-shirt *and* you can wear a *hijab* at the same time, and you can go ahead and mix with boys and sit down with them. The *hijab* may or may not be about segregation. It does not work as simply as people think. You can see all sorts of *hijabs* and headscarves.

The veil and hijab *have become fashion, too. Women wear them in a multitude of different ways in Souk al-Hamidiyeh, where the lingerie is also openly sold. What contact have you had with the lingerie?*
When my wife and I got married, one of her relatives gave us very see-through lingerie as a gift. We've never used stuff like that. Khawla and I lived together before marriage, which isn't very common, so we laughed at it. The woman who gave it is traditional. She wears a veil. She's Islamic, very committed to

religion. Yet the gift was lingerie. In our culture it's the women's role to be sexually attractive to the man and to stimulate his desires.

One of the things Rana Salam and I found interesting is the insistence of lingerie manufacturers that a woman must entertain her husband. According to the Quran: "A woman should dance for her husband."
Yes, dance, whatever. The Quran doesn't say a woman should "dance," and the word "entertain" is not there. The Quran says a woman should be good, faithful, obedient... but "entertain," belly dance—no. That's how the culture has interpreted it.

Thousands of prophetic traditions support these ideas about women. Prophetic traditions are what people wanted to say, and put in the mouth of Muhammad: "It's not us, it's Muhammad."

You're referring to the Hadiths, sayings attributed to the Prophet Muhammad?
No one actually questions the Quran's authenticity except some academic scholars, especially in the West, but there is a question about the Hadiths dating back 1,200 years. Everybody used to question their authenticity, but this became less and less so. Now we have thousands of Hadiths whose authenticity is accepted—you just say it's a Hadith and people accept its authenticity somehow. A lot of Hadiths considered untrustworthy 500 or 1,000 years ago are nowadays considered trustworthy regardless of what scholars say. A lot of scholars consider them trustworthy, too, simply because of the aura of time. Prejudices of the patriarchal culture were legitimized in this manner. People read some positive interpretations in them, but mostly negative ones—especially with regard to sexuality and how women should be treated—as part of the Islamic tradition.

The story is complicated, as usual. Traditional society influences sexuality more than foreign liberal society. It's not the liberal classes in Syria that produce lingerie. It's the most conservative sectors. Their women wear veils. This conservative culture produces sexually explicit lingerie.

Secular modern Western society makes underwear that's very plain compared with Syria's racy variety, where every hole, every embellishment, every sequin, every crochet weave emphasizes women's genitals and erogenous areas.
It's men living out their fantasies. It's probably the same with me when I write about sex. I'm acting out fantasies I couldn't act on when I was the proper age to do it.

In Syria, men work in the lingerie shops, not women. On the other hand, I go with my wife to Victoria's Secret—one of her favorite shops—and I'm one of very few men there. I look at the styles and I tell myself, this is lame. Back home it is simply much, much more.

Syrians have an impressive manufacturing culture, and the manufacturing cities of Damascus and Aleppo are dynamic. They make products that are well received and sold throughout the Gulf.
They're in high demand. There is a little niche. With globalization, we don't have that many opportunities. We have to create these small niches in order to penetrate markets outside Syria. So many people make fabrics now, so it's not about making clothes. It's about making specific items and doing it better. You can penetrate the Gulf market through lingerie. On the other hand, in Syria, we have very-good-quality men's suit manufacturers. There are two or three companies that sell a lot of work in Europe. Many things sold in Europe under one brand or another are actually made in Syria.

Men's tailoring is obviously standardized in the country. In a way, the culture is geared toward men. Interestingly, the trouble with women's wear in Syria is the absence of standard sizing, and this says something about the way the female form remains problematic in a predominantly Muslim society. Your wife likes Victoria's Secret because the bras are designed for comfort. In Syria I saw people fitting bras onto mannequins. Even in this well-developed manufacturing culture, people are still not comfortable with women's bodies.
You might have seen pictures of women wearing lingerie designs, but unfortunately nobody really respects these women. The design process is not made around women's bodies. Despite all this variety and design, the idea of comfort is not there. Khawla, for many years, has not worn a single piece of underwear from Syria. I can get her stuff from abroad, and it's much more comfortable.

Still, Syrian manufacturing continues to do what it does well without any help.
They have to pay bribes. They have to deal with corruption, because the authorities really tax them a lot. They don't get any breaks except the ones they buy through bribes. They bribe their way around laws that are unsuitable for the times, socialist in nature, in order to thrive as capitalist entrepreneurs.

One of the more intriguing aspects of Syrian lingerie is the humor—mobile phones, plastic insects, Tweety Birds—all on thongs. A lot of it is determined by the plastic paraphernalia coming out of China, but the selection of items that appear on G-strings is hilarious. Do Syrians have their own sense of humor?

Syrian humor tends to be raw, especially when it comes to sex. One joke is about a ladies' group that meets on a regular basis, and the women chat about their problems and husbands. One day, a woman in the group is very happy. The other women tell her, "You look chipper. What happened?" She says, "Yesterday my husband, Abu Ali,[5] came in from work. As he changed his clothes, I stuck my hands between his legs and told him, 'Abu Ali, your balls are very cold. Can I warm them up?' It was a night to remember!"

Another woman says, "Well, let me try this with my husband—why not? We haven't been at it for a long time." Next time the group meets, *she's* chipper, and the women ask, "What happened?" She tells them, "My husband came home from work. He was changing his clothes and I stuck my hands between his legs, saying, 'Abu Antar, your balls are very cold, can I warm them up?' It was a night to remember!"

A third woman says, "I will try this with my husband." Next time she comes to the group, she has a black eye and she's walking with a limp. The women are surprised: "What happened to you?" She says, "Well, when Abu Muhammad came in from work and changed his clothes, I put my hands on his balls, and said, 'Hey, Abu Muhammad, why are your balls warm, not like the balls of Abu Ali and Abu Antar?' It was a night to remember...."

In the region, Syrians have a reputation for being earthy and raunchy. Do you agree, or does this reflect the bias of other Arabs against Syrians?

There is some truth to it. Syrian society tackles sexuality head-on and looks at it in a very direct manner, which some people might find strange because it is supposedly a conservative society. There is something Victorian about it. We have this duality. On the one hand, you adhere to certain sexual mores, but underneath it, people enjoy tales of ribaldry. There is overt discussion of sexuality even in gatherings where men and women are together.

Men and women enjoy sexual banter together?

The joke I just told is not something men tell each other or women tell each other.

It is a joke told in a mixed setting. Sometimes it doesn't matter whether the people are religious or not. Sexual jokes are common currency in Syrian society.

Why do you think there is this duality? Is it like the Victorians—on the surface, prim, proper, and pious, while underneath are the dreaded, naked piano legs?[6]
Everybody realizes that traditional mores simply don't work anymore. Let's not forget, Syrians don't get married until their late thirties. So there is no way people are keeping to strict sexual mores. Otherwise everybody would be a virgin at the age of forty! It just doesn't happen, so even veiled women have their illicit liaisons—whether their virginity is maintained or not is a different issue. Sexual contact does take place. People might feel guilty, but nobody can maintain a code that perhaps has never really worked. I think people have realized this to a certain extent.

Is there open discussion in society about topics like homosexuality and premarital sex?
Open discussion, no. These issues feed into the humor. There are jokes about homosexuality, but no open discussion of homosexual issues. There is no acceptance of that.

On the other hand, you do see openly gay people in society. They're not going to come out, but they know you know, and it's accepted because it is a fact of life. People might try to denounce types of sexuality and premarital sex, but everybody realizes that these things go on. So you accommodate these aspects of life. You can condemn homosexuality, but in the final analysis it does exist, and you cannot simply punish everybody who is homosexual.

Also, continuous exposure to what's happening in the world at large, with homosexuality now accepted, makes people in the Middle East more brazen about it. They're not hiding it anymore.

In Aleppo, a friend of mine saw posters advertising a movie featuring the singer Samira Toufic. She and her friends went to see the film in an old cinema. The movie started, then suddenly switched to a pornographic film. Obviously the film or the singer's name was a code. Only a select few knew that a pornographic film was going to be screened. How available is pornography in Syria?
With the Internet, it's more easily available now. Even before, there was always some material, even in theaters—not fashionable movie theaters, but smaller ones. I don't know the code, because I've never been in the scene myself, but I

know that in the theater as you're watching the movie, an operator stands at the gate. If the authorities show up, he presses an alarm that rings in the projection room. They immediately change the reels and put on an Indian film.

In addition, we have casinos and nightclubs. We have clip joints with women from Eastern Europe. In a sense, the taboo of pornography has been broken.

The first time I went to Damascus, in 1971, shortly after Black September in Jordan, I was traveling during Ramadan and the streets were filled with uniformed men.[7] *Syria then appeared to be highly militarized and socialist, which made it seem less decadent than the West. Now when I visit, I pick up telltale signs in the hotels that Damascus is a city at play.*

I was born in 1966, so I remember the early 1970s vaguely. I remember, for instance, my parents would take me to nightclubs. At that time, nightclubs were a family affair. You could go and see the dancing. You could dance yourself at the disco and have something to drink—that was not a problem—and there were shows. So, really, in the early 1970s Damascus was not that innocent. What happened in 1975, 1977, and onward, the clashes with the Muslim Brotherhood, created a very conservative environment. This kind of social life died down. It slowly started to emerge again in the 1990s, and right now you can see, once more, nightclubs and bars downtown.

The "old" Damascus is experiencing a revival—a return to raunchiness. Once again, the city is being true to its contradictory self. We are not a homogenous society. The modernization of Syria since the early twentieth century is a constant challenge to the layer of homogeneity that traditionalists want to hold on to. It is a surfacing of the contradictions.

Taking off the veil is synonymous with taking off this veneer under which our contradictions were hidden for so many centuries. Modernity is the process of removing the fig leaf from our contradictions, and frankly it suits us better. We are much livelier when we are more open.

I was told that Damascus has become a new sex capital for Arabs from the Gulf. There are nightclubs in the mountains just for them.

The interesting thing is that they might bring their families with them to Damascus, and they might leave them at home in Damascus, but then they go to the nightclubs. Sex tourism hit Syria in the mid-1990s, and it's ongoing.

"Sex tourism" has sleazy connotations. Yet when I take out our collection of Syrian lingerie and look at the outfits that women who dance for their husbands wear, or the flowered bra-and-thong set that plays "Old MacDonald," it's not sleazy. It's playful—not childlike, but related to child's play.

We're a culture of big kids when it comes to sexuality. We're still exploring with the fascination of teenagers. But there is also an element of cynicism about all these issues because guilt, in the final analysis, is involved. We're not comfortable with our sexuality. We have not attained peace of mind. No one is willing to reject traditional mores completely. In practice they do, but not in theory.

Photographer Omar al-Moutem showed Rana and me photographs of Eastern European models modeling lingerie. There were styles with PVC masks, bras with zippers, and little whips. Omar explained that these outfits were for people who like to hurt other people when they're having sex. He felt the need to explain it simply and slowly to us because he thought that as "respectable" women we would never before have encountered S&M. We found that very endearing. It said something about his innocence.

In Syria there is no awareness about how Europeans have become blasé about these issues. People in the West understand different aspects of sexuality. In the Middle East we're still coming to terms with them, at least openly. Secretly it has happened, but we do not talk openly about it. We're still sort of fascinated.

A cartoon by Syrian cartoonist Ali Farzat shows a torture chamber filled with implements of cruelty—pulleys, manacles with hooks and thorns.[8] An unconscious man tied to a wall has obviously been tortured, his severed hand and foot tossed on the floor beside a bloodstained saw and wrench. The torturer sits on the floor of the cell, watching TV. Onscreen, a man and a woman embrace; a single tear runs down each of their cheeks. A red heart floats above their heads. The torturer, too, is crying. The cartoon suggests that sexuality and romance are diversions from the more brutal aspects of Syrian life.

In a sense, they are a haven. We're going through a very tough period, and we've been living under a very cruel regime for a long time now. Storytelling, soap operas, and, right now, this return of raunchiness and nightclubs are an escape from the insanity around us—not only in Syria but in the region as a whole.

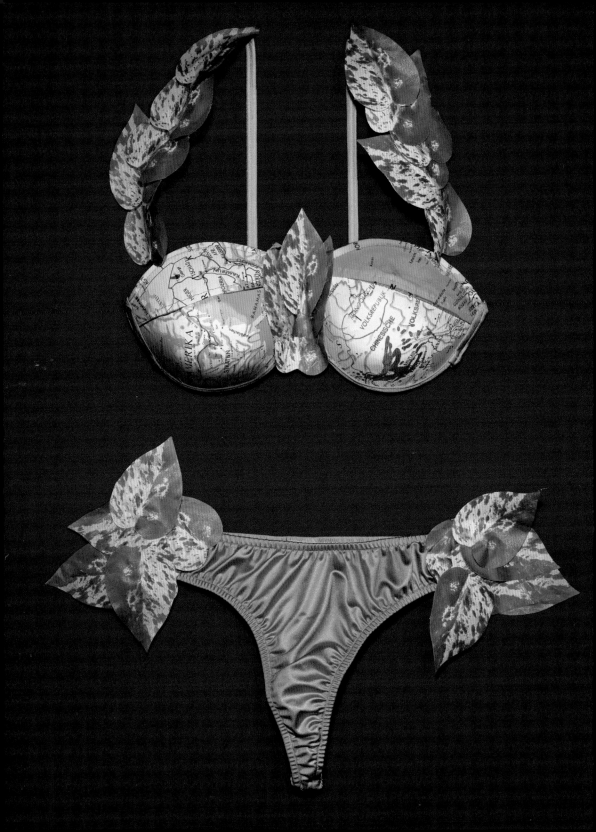

Photographs by Gilbert Hage
Up Close:
Intimate Still Lifes

Gilbert Hage is a meticulous, inventive artist. "My objective," he writes, "is not based on taste and aesthetic but on the tactics of representation. I am thoroughly convinced that the human face or the portrait is neither determined nor real; however, it is receptive, flexible and subject to changes. It oscillates between realities and illusions."[1] His still-life photographs of Syrian lingerie represent the illusion of sexual freedom in a country not known for its openness about sex. At the same time, the playful nature of lingerie design suggests confidence and exuberance in all matters sexual. Audacity and anxiety exist in the same place at the same time.

Interviews with Syrian Women
Inside Story

The women working in the industry or shopping in the lingerie stores in the souk were the most forthcoming. Others who were not interviewed in factories or stores appeared, at times, to be nervous about entering into any kind of discussion. Many refused to talk to us at all or became angry when they saw product shots or examples of the lingerie. They often thought that it was somehow demeaning for women to reveal their private lives, that by making a comment they were impugning their own respectability or the status of Arab women in general. Hassan Ramadan, a translator and researcher, has described this attitude as "chastity of the mind."

However, there was also another very understandable reason voiced by a woman in her sixties. At a time when Syria is much maligned in Western media the subject could be open to negative interpretations. She said, "Damascenes are proud of their city and culture. No one will talk to you."

Those who did, work in government offices and schools; others come from the arts—fashion, music, and literature in Syria and abroad. Some felt more comfortable speaking under a pseudonym or by withdrawing their names. The deepest personal insights come from the women interviewed by Eugenie Dolberg, a photographer and social activist.

With the continuing debate about Islam and gender, these are the rare voices that should be heard.

—The Editors

The Secret Life of Syrian Lingerie

"It was my birthday and I met some friends in a café in the old city of Damascus. We sat drinking coffee and smoking hubbley bubblys. My friends brought me some gifts. I started happily unwrapping them. One friend had brought me a bag, another friend, a beautiful little box. When I opened the box I was so surprised, there was absolutely incredible lingerie inside with a bird with real feathers in a little nest. When you turned the bird on, it lit up and started singing. Everybody in the café was laughing and I felt so shy. I wanted to kill my friend! I took my presents home and made sure I hid the lingerie very carefully. I completely forgot about it. Months later, I was away in my village and my family was redecorating the house. I received a phone call from my sister who told me to come straight back to Damascus. I was so worried—during the whole trip I had no idea what had happened.

My mom found the underwear! She was so angry with me and asked me if this was how I repaid her for the freedom she had given me? I tried to explain but she was too angry. The only reason I didn't get into real trouble was that she couldn't tell my father. Even so, she kept a much sharper eye on me for months."

— *Name Withheld*

"Lingerie is such an embarrassing issue for Syrian women to talk about, and that's probably why it's called 'the secret life.' Since I'm not married, lingerie doesn't mean anything to me. When I choose my underwear I look for good-quality cotton, one that doesn't cause allergies. Frankly speaking, I'm not interested in spending money on lingerie. I prefer to buy brand names like Mango, Stefanel, Guess, et cetera, rather than buying something nobody will see. Are you surprised? In my opinion, it's a matter of taste or preference."

— *Elsy Violin (a pseudonym)*

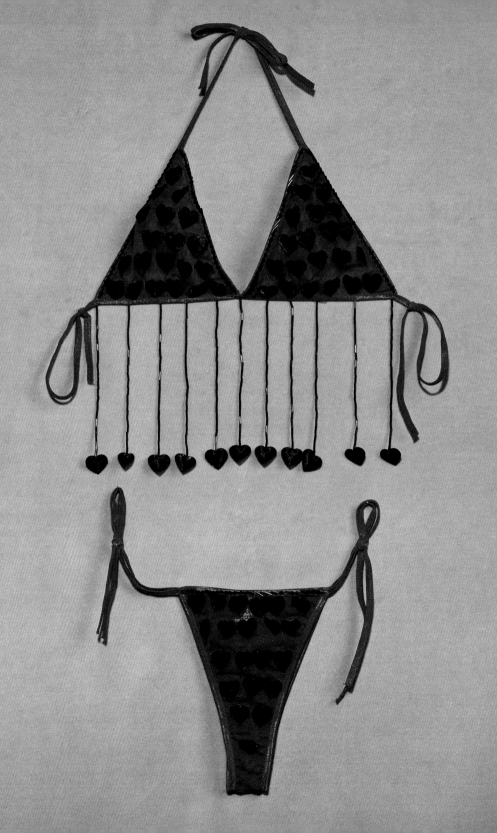

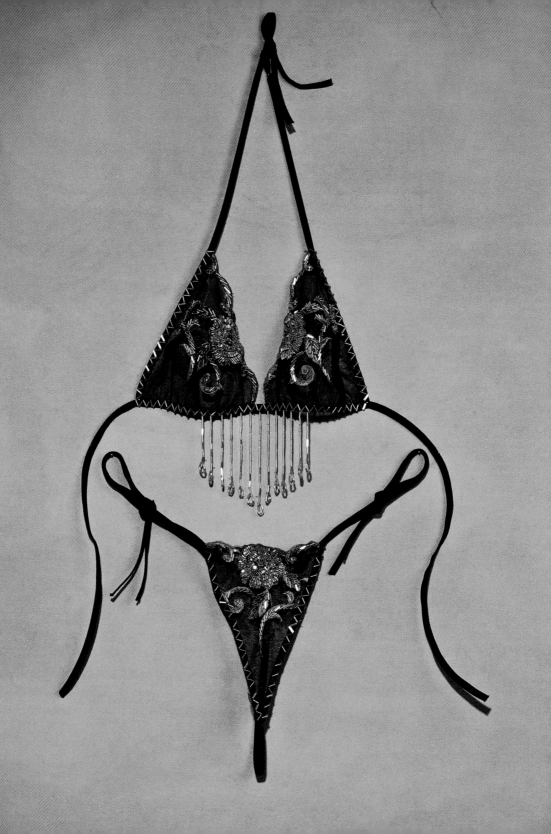

Photographs by Gilbert HageInterviews with Syrian Women

"In Damascus, I was looking for someone to make beaded thongs for Karizma, my lingerie business in London. When I explained to people in the souk what I wanted, they kept showing me standard underwear. Finally a woman I was talking to explained, 'You're looking for sexy underwear.' She gave me the name of the guy who made fantasy stuff, and I made an appointment to go and see him. A veiled woman opened the door, invited us in, and offered us dates from Medina. The man was running a bit late, and she said to take a look around the showroom. It was filled with the dodgiest things I had ever seen—chains, hardcore leather S&M, and stuff like that. As time passed, I kept thinking, 'Am I in the wrong place?' I was with my mother, my mother's cousins who are very conservative, and their wives who are even more conservative—all of them veiled. Finally a very pleasant man arrived. He asked me if I had looked around and if I saw anything I liked. I explained my idea to him. He understood straightaway and showed me exactly what I was looking for. In the meantime he showed me other materials as well. He had rubber you could stretch that would change color according to body temperature. These were different things in lingerie design that never occurred to me."

— *Zina Azem*

"One day a friend of mine was having problems with her husband in the bedroom—if you know what I mean. She went to the pharmacy to get some drugs to help him. When she told the pharmacist, he said to go and buy some underwear in Souk al-Hamidiyeh. He said that, as a man, he thought her husband might feel his masculinity insulted if she came to him with drugs, and it might make their problem worse in the long run. My friend was a little shocked but thought she would try the pharmacist's advice. She bought some amazing pink underwear shaped like a rabbit, you pulled the rabbit's ears and it lit up, popped open, and sang a song. It worked a treat."

— *Name Withheld*

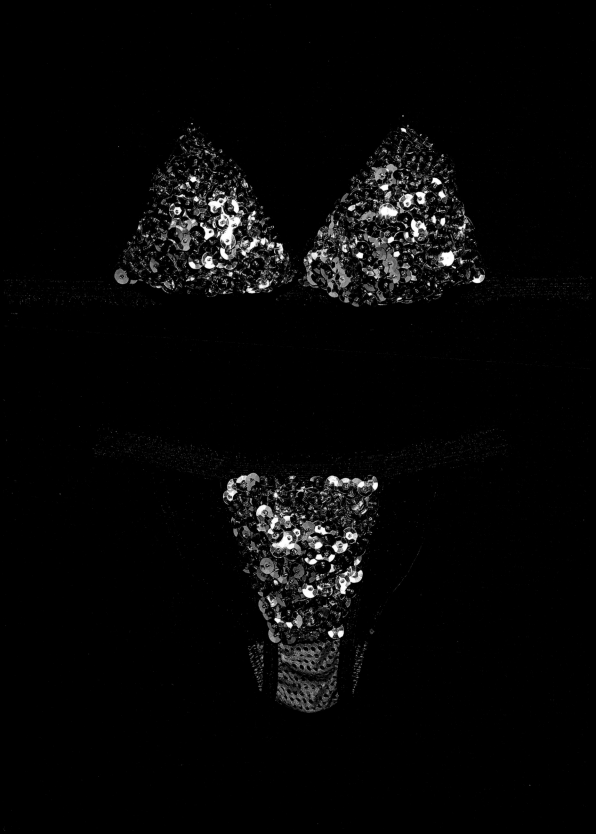

Photographs by Gilbert Hage   Interviews with Syrian Women

"What's different about this lingerie is that it is linked to a small category of people. There are subcultures in Syria, and subcultures everywhere else in the world that would be interested in this kind of lingerie, starting from Le Lido or Moulin Rouge. What's also interesting is that the lingerie is, again, a very small part of fashion when compared to the greater Syrian ready-to-wear lingerie produced by companies like Hanin, Asseel, and Flora. These women's lingerie firms are the country's equivalent to Hanro, DKNY, or La Perla for normal women.

In the early 1990s, after the civil war in Beirut, there were increasing commercial ties between Syria and Lebanon, and Syrian women bought much of their comfortable wearable lingerie from there. This exchange not only educated the Syrian middle- and upper-middle-class consumer, it was also an incentive for Syrian companies who wanted to manufacture for this market at home. We have a thriving modern industry and nowadays bras are cut by lasers for $40 instead of $400. In ladies' lingerie, people are very serious about what they make and do—whether they grade and size patterns or manufacture well-finished cotton camisoles. There is a level of what I would call 'whorishness' in a lot of products in fashion in general, and in lingerie specifically, that's not sexy or sophisticated. Some of the sexy Syrian lingerie comes out of a naïveté, as if the designers and manufacturers haven't been exposed enough to comfortable women's styles. It is also born out of social frustrations typical of a society in transition and flux. Syria is a country in transformation, and it is indeed an extremely difficult balance to keep. Within just a few years, satellite TV has made all manner of things available and possible. People are more exposed to the rest of the world, and there is much more to choose from. Yet the social structure, like in any other country, is far more layered, complex, and sophisticated than what many in the so-called West portray, and in some cases portray to develop their own divisions and categorizations. Then again I have women friends who plan to pursue advance degrees in medicine, fashion, as well as linguistic or religious studies abroad because they're aware of the challenges of the new future world. I have male friends who have worked as clothing models in the past, had live-in foreign girlfriends, and are now returning to their family's traditional roots. It's not right to generalize with such a complex country, as there are countless social trends as well as individual cases occurring in Syria—some follow a harmonious path and others rub up against each other and create friction."

— *Taleen Temizian*

The Secret Life of Syrian Lingerie

"Until I was twenty years old, lingerie was out of my dictionary. Then some friends of mine were getting married and they asked me to bring them lingerie during my trips to the United States. So I grew to love buying lingerie, my favorite place is Victoria's Secret. I love their styles. They are simply very sexy. The baby doll styles are really hilarious and the bridal collection is superb. I bought some lingerie for myself. I don't wear it now, but I cannot wait to get married to try it on. When it comes to Syrian lingerie, I love Flora. Most of the salespersons are women. I hate the fact that men sell lingerie in Syria. It ruins the shopping experience for me. I feel shy and embarrassed, yet there are nice designs of Syrian lingerie, except for the feathered ones."

— *Lilian (a pseudonym)*

"The underwear makes me laugh so much. It is not sexy at all! At least, I don't think so. It is so old-fashioned and mirrors the very old-fashioned ideas about sexuality in the less educated classes of Syria. They are trying to make it terribly modern and use all the modern electrical gadgets one can imagine! The mind boggles!"

— *Name Withheld*

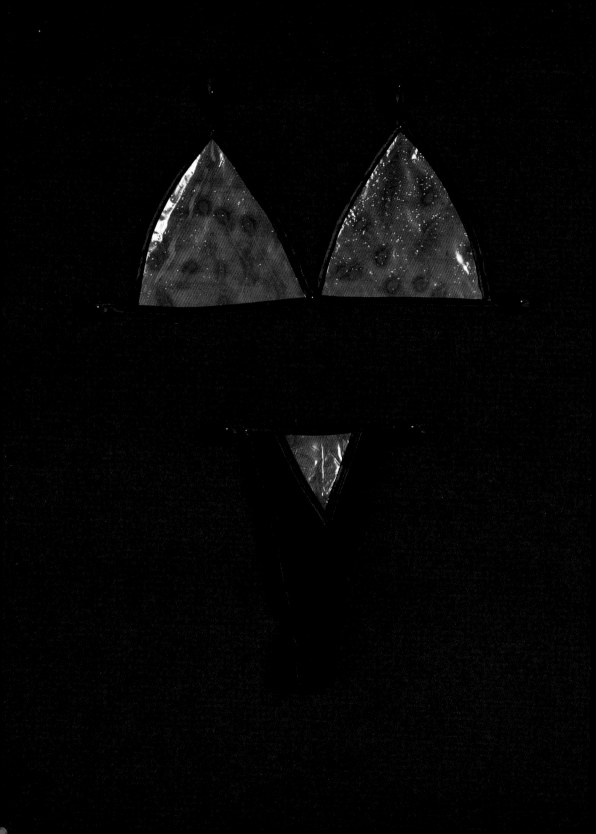

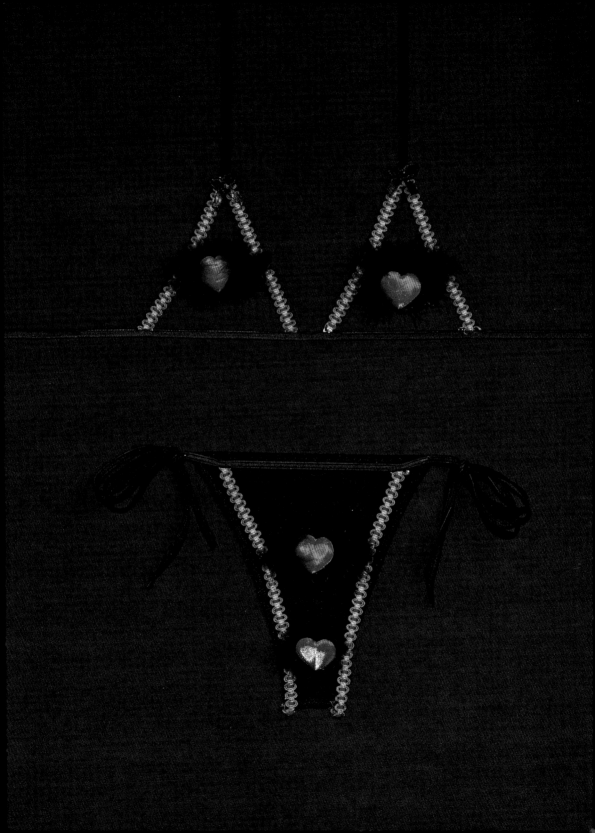

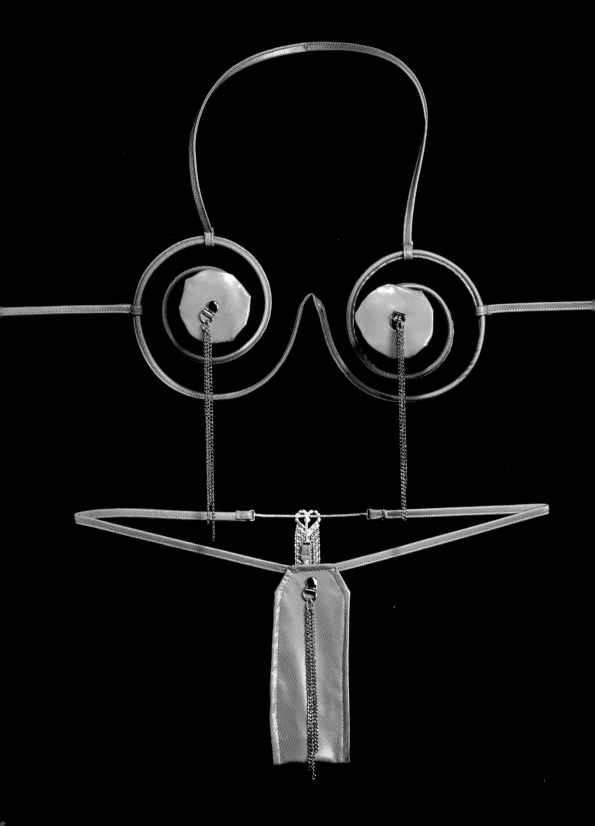

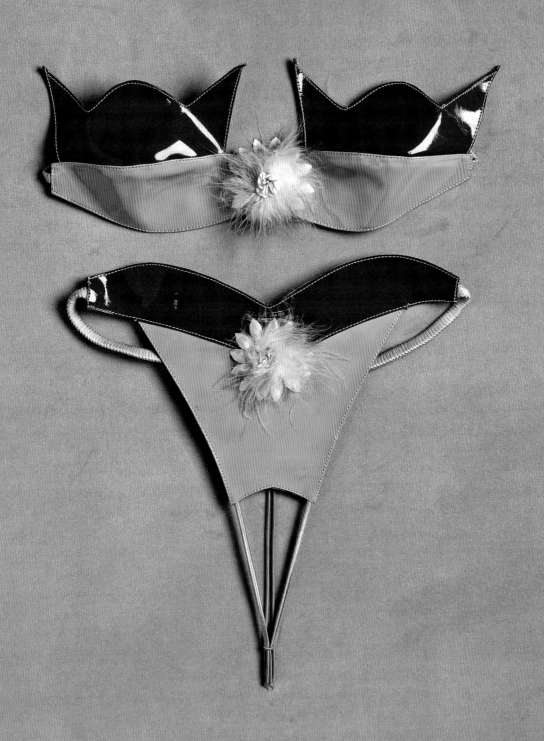

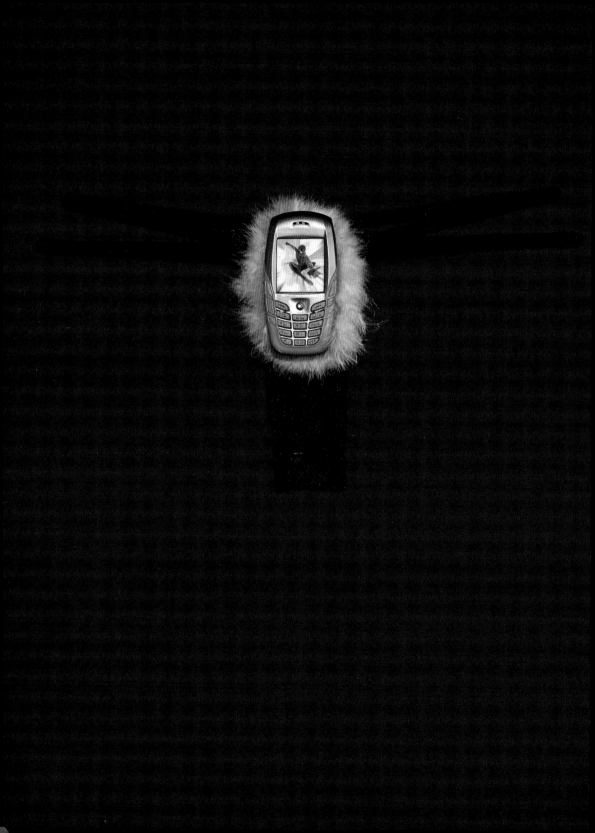

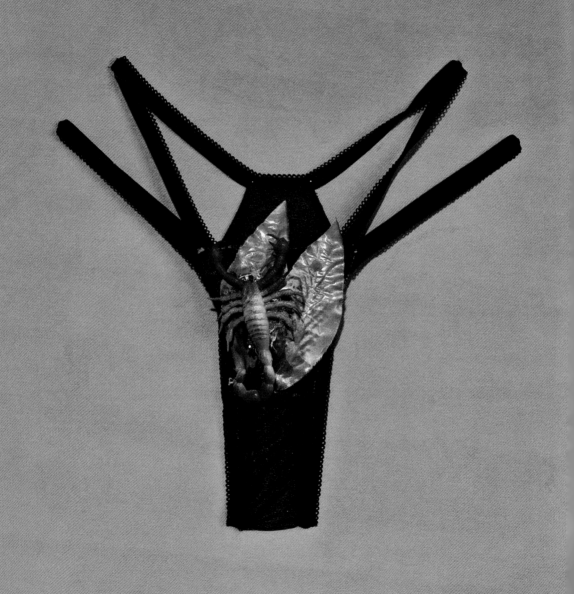

Photographs by Gilbert Hage   Interviews with Syrian Women

"I asked my Iraqi Christian friend about the underwear. She has been living in Syria ever since her family was forced to leave Iraq by a militia. She laughed a lot and said it was a crazy idea to wear 'that' underwear but being crazy herself she would like to one day try it. Unfortunately she didn't think her husband would be into it. In fact she thought he might go so far as to kick her out. He would think she had completely lost her marbles and would feel threatened by her sexually aggressive behavior. We laughed so much! She said it came down to courage. She had the courage to try new things but he didn't!

My Muslim friend, on the other hand, thinks this underwear is completely normal, and would happily wear it to please her husband. Honestly, I was so surprised when I found out that a lot of her friends wear it, too. It is all about having fun sex, and they like it! I work in a government office and sometimes some of the girls bring the underwear to work. Oh my God, it makes me so shy, my face burns up red and I have to close my eyes. They show it off and have a good laugh. The same girls are always giving advice on how to have a good sex life, pleasure their husbands and even themselves! Especially before girls get married. They are very open but I cringe with embarrassment. They really want to show everybody that they have a good sex life. I don't think it is about freedom. They want people to know they can pleasure their husbands. My friend said to me that she even dyed her pubic hair to keep her husband happy at home and away from other women.

Before this interview I hadn't really thought about it, but actually it is really interesting and says a lot about Syrian society, although of course my knowledge just comes from a small group of my friends. It is absolutely not something that is publicly talked about. For Christians, this underwear is nothing but embarrassing, old-fashioned fun; for Muslims it is something very normal—they not only accept it but also enjoy it. The more religious an area is, the more risqué the underwear becomes. I think that Muslim women have less freedom on the outside so to compensate they have more freedom on the inside. It is exactly the opposite for Christians. Where Muslim women would rarely say no to her husband in the bedroom, I can. Often I say no if something is embarrassing and I think I have missed out on some really pleasurable parts of making love. Now the tone of my relationship is set and it would be weird to break it.

I love sex but nobody in my community talks about it. Can you imagine, I find it even difficult to look at the underwear in the shops? After this conversation, I am curious, maybe I will give it a go after all."

— *Name Withheld*

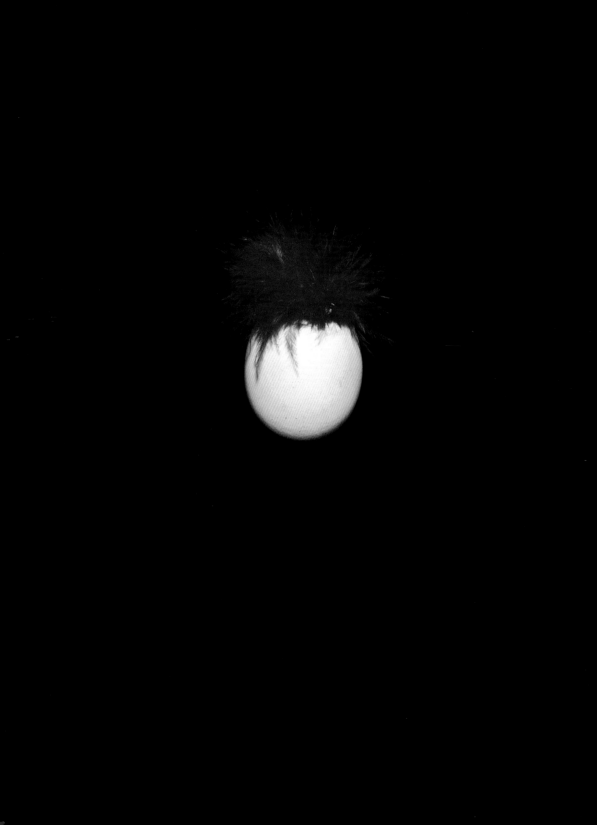

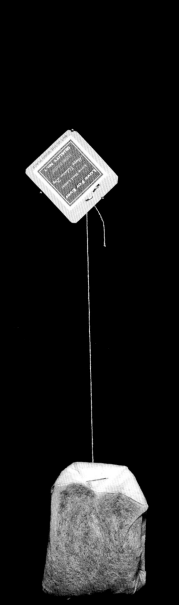

"I was brought up in a strict middle-class family from Aleppo in northern Syria. I never saw my mother or my sisters buy lingerie. We were taught to be good wives in ways other than seducing husbands—housework, nursing, and understanding. We were not allowed to talk about sex; we only used to hear secret jokes about it. In Syria I saw the lingerie hanging on the fronts of shops, and I remember how I use to feel—disgusted. Now in London I still see this kind of lingerie being sold, but I understand that people everywhere have similar experiences when it comes to the body."

— *Ghalia Kabbani*

"My experience was a lucky one because I never bought lingerie from shops that sell extravagant lingerie. By the time I started buying lingerie, many Syrian companies had already started manufacturing simple, beautiful European styles. The photos in the book are a bit of a surprise to me. I used to see lingerie similar to this in very specific shops, and they always made me smile. Women who wear such lingerie have a different background than mine. They buy styles, thinking that they make them look sexually attractive to their partners. They probably care less about their comfort or personal satisfaction with the lingerie. The purpose of lingerie for many individuals, including me, my friends, and family, is comfort first and attractiveness second. Another type is simple and sexy. Women always need to feel attractive. Ultimately my own favorite is simply designed, elegant, and sexy. I would never wear the kind in the photos."

— *Lina Makss Elias*

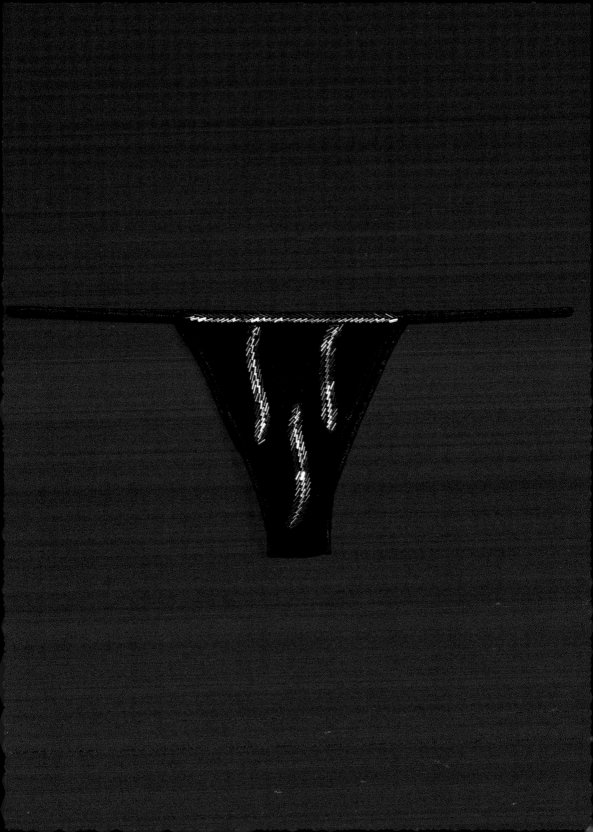

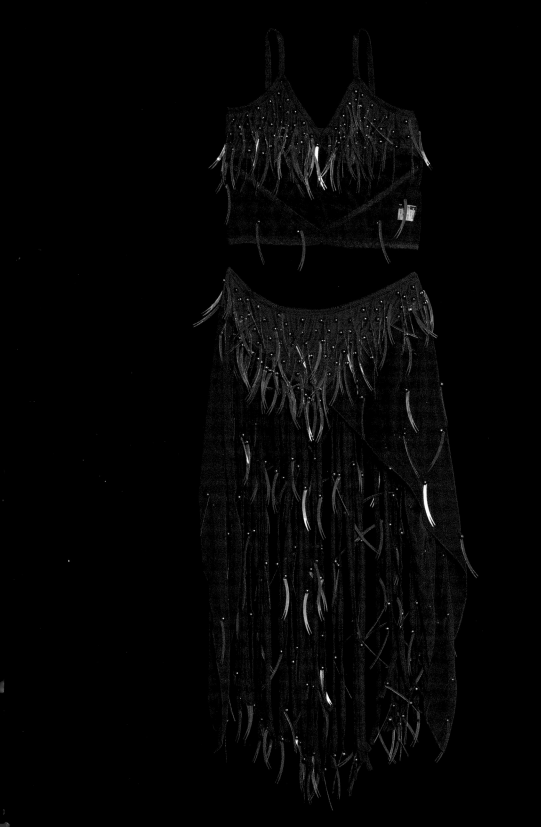

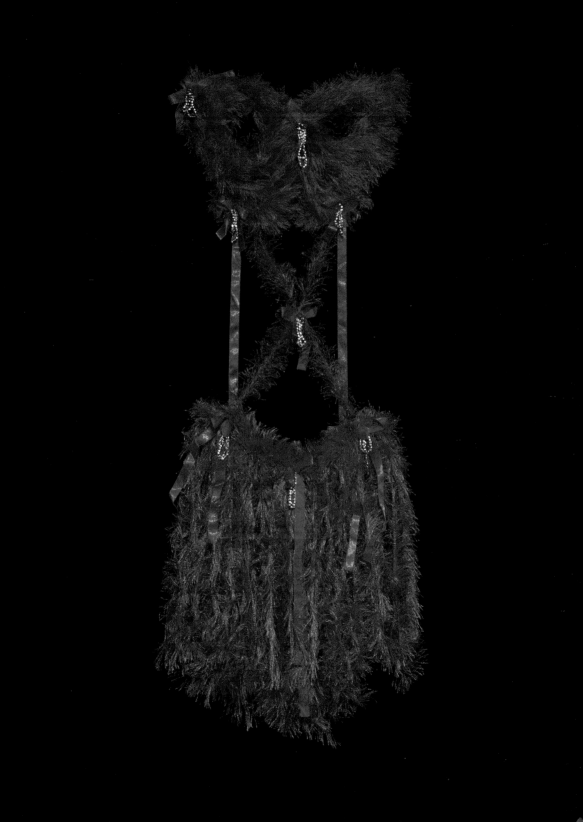

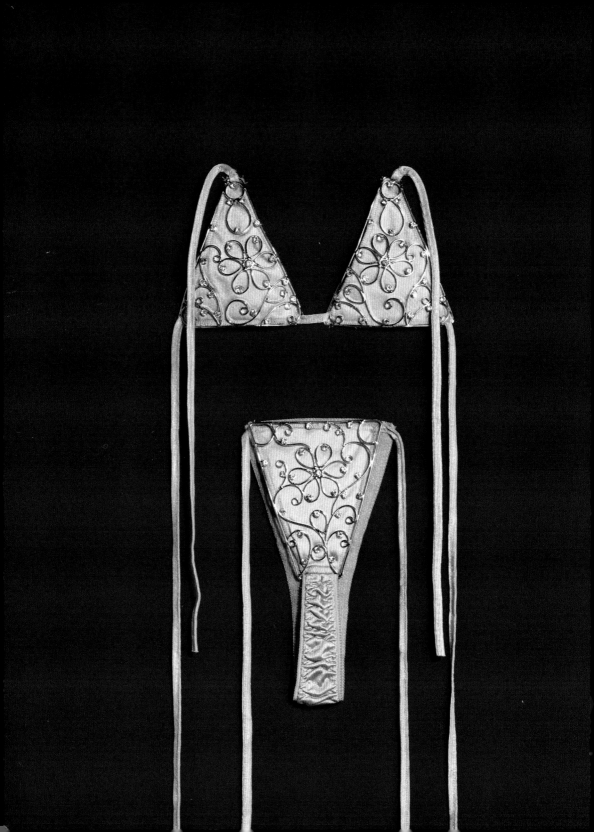

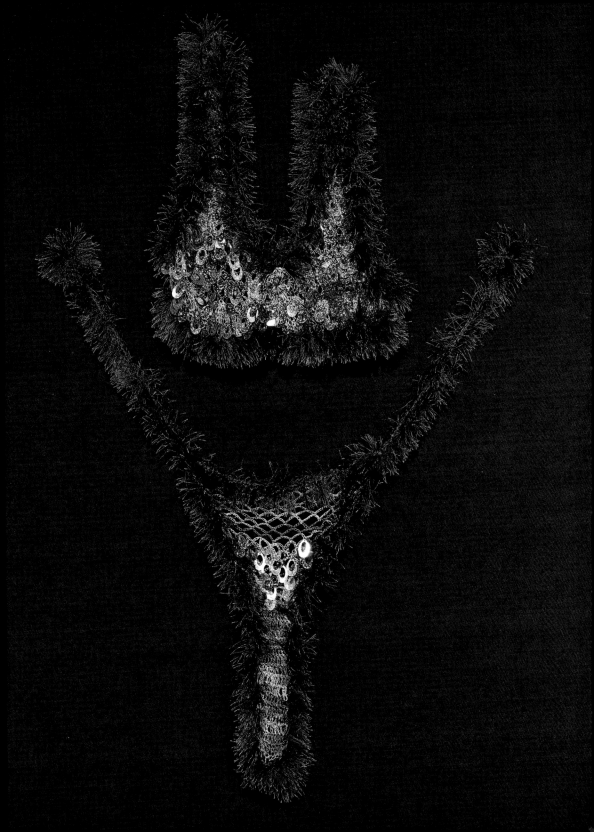

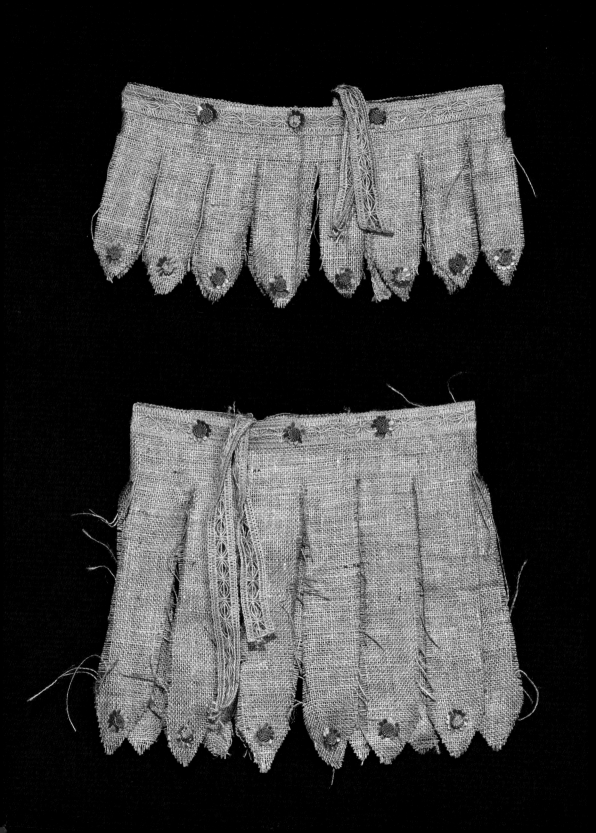

The Secret Life of Syrian Lingerie

"I can tell you that lingerie means a lot for some women in Syria while it means nothing to others. Some Syrian women think that buying lingerie is a waste of money and it has no role in their sexual lives. While for some traditional people, it is a must to buy lingerie with all different colors and styles before getting married. But this doesn't include only lingerie; it also includes all kinds of clothing. I think that not being able to wear what you want in daily life is a factor, which makes some girls feel that only marriage will give them the chance to show what they've been hiding. Personally, I am not on any extreme side. But I don't like the idea of looking at lingerie displayed in shop windows. None of the styles are what I like to buy."

— *Name withheld*

"I've never bought Syrian lingerie myself, but as a fashion designer, I've studied it as part of my work. The first time I noticed lingerie was during the Motex Fair for Syrian prêt-à-porter in Damascus. There were two sections: women's lingerie and children's wear—as if all that we have to do in Syria is wear sexy clothes so we can have children. But before I continue, we need to make two points. First, women in Syria live—in general—in good conditions: they go to universities, work, and make their own decisions in life. The second is the nightlife. They're like other women: seduction is fun. It's not a bad idea to think about seducing a man to keep him around—the first lesson old ladies give to a girl on her wedding day. It's the same everywhere. The only difference is that the Syrian version of seduction is more intimate than that in *Sex and the City*."

— *Layla Basha*

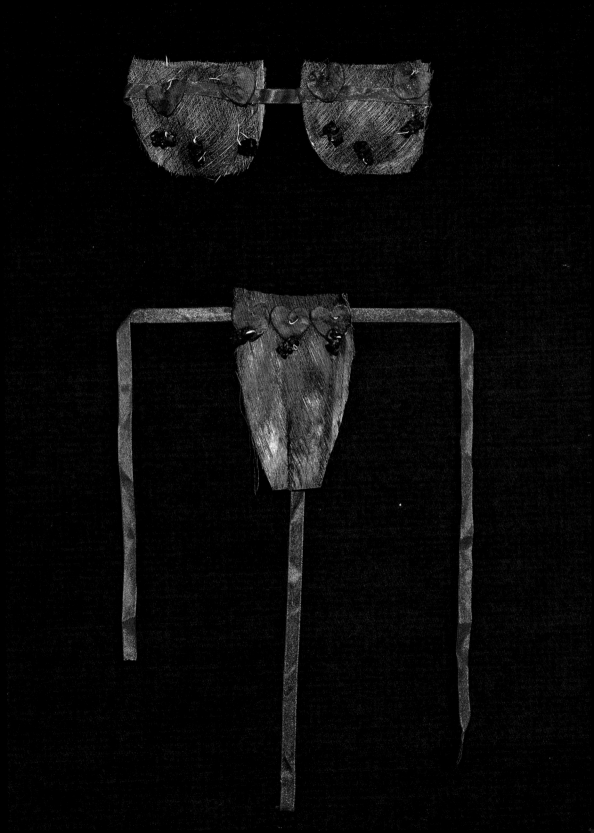

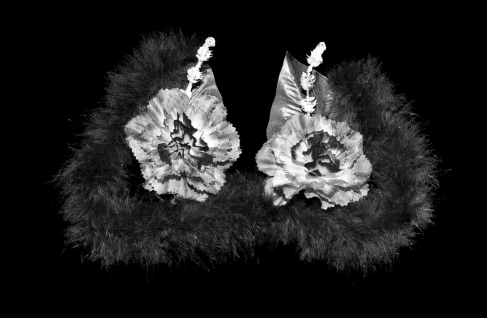

Photographs by Gilbert Hage Interviews with Syrian Women

"The lingerie is almost done in a naïve sweet innocent way. It's not sick or perverted. Even the sadomasochistic ones are soft; they don't have a dark, pornographic side to them. It's something to have fun with, to enjoy.

When I went to the main lingerie shop in Souk al-Hamidiyeh, I didn't feel it was peculiar or out of the ordinary. The shop owner was relaxed and smiling and didn't give the sense he was embarrassed that he was selling all this crazy stuff, which he said he created. He kept bringing out so many different styles. I felt I was in a candy or toy store.

Although Middle Eastern society expects a woman to be a virgin and get married—the goody-two-shoes, all saintly in society on the outside—on the inside, people are doing everything they want. So the lingerie is a contradiction. You can't discuss it in public but privately it's a different matter.

I'll conclude with a story from my friend. At the office where she used to work, there was a cleaning lady and my friend asked her what she thought of the lingerie. The woman said, 'I love it. I have the one with the mobile phone. I put that one on'—and this woman is veiled by the way—'when I want to have sex.' Then she sits on the bed, naked accept for the mobile phone thong, calling to her husband, 'Dring, dring, dring, dring, are you going to answer the phone?'

These creations make it easier for a couple to talk about sex, a topic that is a huge taboo in the Middle East. Suddenly it's not so heavy and serious. They also get a chance to laugh a lot."

— *Lena Lahham*

Photographs by Omar al-Moutem, Amasel Fashion, and Angel Lady

Modeling Lingerie: Product Photography from Lingerie Manufacturers

Omar al-Moutem was taught photography at home by his brother. When he started working in a lingerie shop, it was only a matter of time before he set up his own studio. To find models, he would send an English-speaking friend to wait outside the clubs and hotels of Damascus and ask the Western women working there if they wanted to model lingerie. Albeena was one of al-Moutem's more popular models who posed for all the local lingerie companies. She was from Russia, Bulgaria, or Romania. Like many of the women he worked with, al-Moutem didn't know much about her.

Some of the women were just passing through, and after a few sessions, they asked the photographer to take pictures of them—not smiling sweetly as they did for the lingerie albums. They wanted to pose with sexy, pouting expressions more in line with their next career move to Italy. Other models made Syria their home, and one even married and settled down. (Now, when she models lingerie, she disguises herself by wearing wigs and sunglasses.) In the photographs, by Omar al-Moutem and those provided by Angel Lady (page 128) and Amasel Fashion (pages 130–133), the models look innocent, friendly, or sometimes dazed. It could be the late nights. Posing and earning pocket money from the lingerie companies are much less of a hassle than working in the cabarets.

In a predominately Muslim country, these photographs of women's scantily clad bodies, widely distributed and openly displayed in the souk, represent a fashion industry. On the part of the companies, it has been an effective grassroots marketing campaign. Not only men look at these photos; mothers, mothers-in-law, and brides-to-be appraise them, too. But lingerie has its seasons, as do the models. As more panties, thongs, and novelty sets are exported to countries more religiously conservative than Syria, selling tactics change. Today's up-and-coming Syrian lingerie model isn't Eastern European. She isn't alive and kicking. She's a mannequin nicknamed "Suzy."

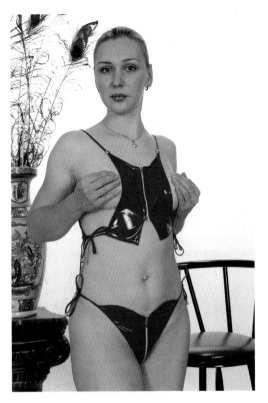

لانجري لور

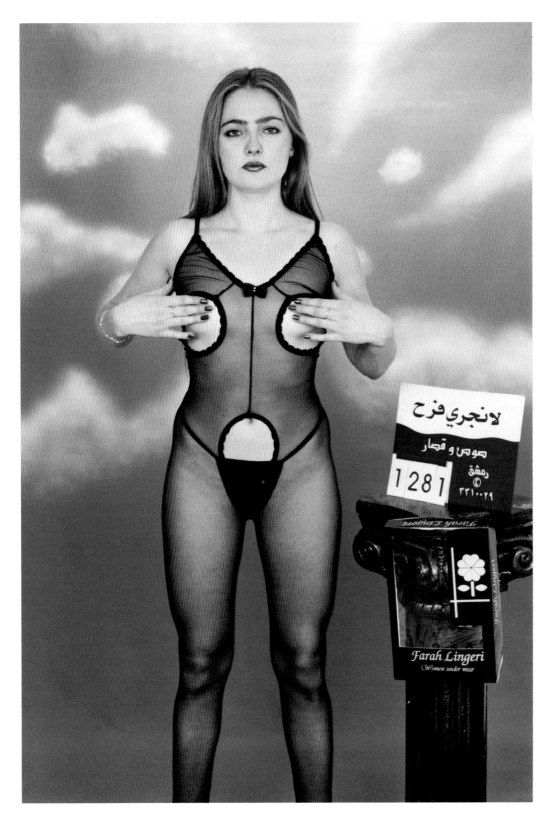

Poetry and Self-Portraiture by Iman Ibrahim

Coda:
A Room of One's Own

I'm training to be human. I document my life by writing in my diary and expressing my emotions through poetry. I corroborate my life steps through photography, because a photograph is a soul. I'm not fond of beauty, because I don't care to be beautiful in my photos. Photography is a way of expressing myself, like poetry.

It's not that I don't respect the ideas of other people. But in the company of others, you can't always be yourself. I like to see myself through my own viewpoint. By showing details to myself, I find things I didn't know about myself.

I don't want my life to pass without me.

The Secret Life of Syrian Lingerie

Untitled

Darkness is
Not seeing my hand
Fear is
Seeing it
And not trusting it

Poetry and Self-Portraiture by Iman Ibrahim

Ray

Every thread
I tie to a tree branch
Will be a ray
That guides me towards an unknown path

Dream woodcutter

I finally know
How to protect my fragility
I know how to be
A dream woodcutter

Poetry and Self-Portraiture by Iman Ibrahim

Untitled

The virgin reclining on her dreams
Sips her future
Until drunk

The Secret Life of Syrian Lingerie

The dressmaker

Brothers, learn
How to wear shirts of tenderness
Not to remain
The rest of your lives
Naked and lonely
Without a sister
Who devotes her life
To sewing your happiness

Poetry and Self-Portraiture by Iman Ibrahim

The blaze of love

Me . . . like the start of every spring
Changeable
My voice a primitive trap
Where only monsters fall
My first gasp
Rocks my final peace of mind
Red garlands on my underwear . . .
Is this the blaze of love?

The Secret Life of Syrian Lingerie

Untitled

All the whores of the universe
Will enter heaven
By divine pardon
As for me
I will remain pure as a saint
Who, exhausted from crawling,
Has chosen Hell

Poetry and Self-Portraiture by Iman Ibrahim

Untitled

My neck gets shorter . . .
As my bedroom ceiling bears down on it
Humped like a kneecap
Yet my heart is brimful

Feathers

In the cages of a dream
My father finds only feathers
As he awakens to confiscate the dream
It visits me
In my sleep
. . . but myself

Poetry and Self-Portraiture by Iman Ibrahim

Untitled

Time passes
Fast like sin
Slowly like regret

Footnotes

Introduction:

1. In 2006, Egyptian cleric Rashad Khalil, a former dean of Al-Azhar University's faculty of *shari'a* (Islamic law), issued a fatwa against nakedness during intercourse, a state of undress that effectively annuls marriage. While other scholars dismissed the ruling, Al-Azhar's fatwa committee chairman, Abdullah Megawar, argued that married couples should not look at each other's genitalia and suggested the use of a blanket during sex.

Damascus: From Factory to Souk

1. *Sexy Souks,* at Point Éphémère from March 8 to April 1, was organized by Maison des Cultures du Monde in Paris.

Competing Thongs:
The Lingerie Culture of Syria

1. During the Baghdad Summit of 1978, Saudi Arabia and other Gulf countries pledged monies to Syria as a front-line state in the Arab-Israeli conflict. "Syria," *Country Studies Series,* Federal Research Division, Library of Congress. http://www.country-data.com/cqi-bin/query/r=13538.html.

2. Megan K. Stack, "Art of the Impolitic in Syria," *Los Angeles Times,* August 3, 2004.

3. Observation made by the Lebanese fine artist and photographer Gilbert Hage in his discussions about photographing lingerie for *The Secret Life of Syrian Lingerie.*

On Love and Lingerie

1. In Damascus, I saw women wearing all of the following: burkas, chadors, veils, *hijabs,* *jilbabs,* and *abayas.* In my essay and film, I call the general covering most women wear burka and veil, because veil is not descriptive enough. The burka in Damascus is a veil that covers the face and entire head but has a cutout for the eyes, as opposed to the full burka or Afghan burka, which covers the entire body and has a grille over the face through which a woman looks. The eyes are covered with a net curtain allowing the woman to see out but preventing other people from seeing her eyes. The Afghan burka may also have slits for the hands. Both kinds of burka are used by some Muslim women as an interpretation of the *hijab* dress code, i.e., the practice of dressing modestly.

For women, the *hijab* is a type of head covering. A square of fabric is folded into a triangle and then placed over the head and fastened under the chin. This is probably the most common current usage. The *boushiya* is a veil that is tied on at the forehead and falls to cover the entire face but has no cutout for the eyes; instead, the fabric is sheer enough to be seen through. The chador, a traditional Iranian outer garment covering the head and body, is a full-length semicircle of fabric that comes to the ground. It does not have slits for the hands and is held shut with the hands or teeth or is simply wrapped under the arms. The *jilbab* is a type of outer garment that looks like a long raincoat or trench coat. The *niqab* is any type of veil for the face or outfit that covers the face. The half *niqab* is a veil that is tied on at the bridge of the nose and falls to cover the lower face. The full *niqab* is a veil that is tied on at the forehead and falls to cover the entire face; it has a cutout for the eyes.

The *abaya,* an overgarment, is the traditional form of dressing modestly. Traditional *abayas* are black and may be a large square of fabric draped from the shoulders or head, or a long black caftan. The *abaya* should cover the whole body except the face, feet, and hands. It can be worn with the *niqab,* a face veil covering all but the eyes.

2. "The Second Wife," translated by Elizabeth Warnock Fernea, is a song sung by Berber women recorded from an oral recital in the 1920s by Mirida Aït Attik in the Tachelhaït dialect by the writer, poet, and painter René Euloge in *Les chants de la Tassaout* (Casablanca: Maroc Editions, 1972).

3. This is a well-known song by the famous Egyptian singer Farid al-Atrash.

4. The taboo is slowly fading away in certain countries. Gradually, belly dancing is becoming accepted as an art form. A few successful belly dancers are respected artists who perform in Lebanon and elsewhere in the Middle East.

A Fundamentalist Changes His Mind: Sexuality and Humor in Syria

1. Noura Boustany, "A Modernizer Challenges Syria's Old Order," *Washington Post,* July 30, 2004; and Lee Smith, "The Way We Live Now: 2–13–05; ENCOUNTER; A Liberal in Damascus," *New York Times Magazine,* February 13, 2005.

2. *Menstruation* was published by Saqi Books in the United Kingdom and by St. Martins Press in the United States, in 2001.

3. The Web sites for DarEmar and Al-Tharwa Project are at www.daremar.org and www.tharwaproject.org. Amarji—A Heretic's Blog is at www.amarji.org and http://amarji.blogspot.com.

4. Sayyid Qutb (1906–66) is considered the father of modern Islamic fundamentalism. See Robert Irwin, "Did This Man Inspire Bin Laden?" *The Guardian,* November 1, 2001.

5. *Umm* and *Abu* mean "Mother" and "Father," respectively. The name that follows in an informal appellation is usually that of the first-born son; thus Abu Ali is "Father of Ali."

6. The Victorian habit of covering furniture legs with skirts, ostensibly to protect against brooms and general wear and tear, has given rise to stories about Victorians' exaggerated sense of modesty and morality.

7. After violent clashes, the Jordanian army expelled all Palestinian guerrillas from the country by July 1971. See Peter Mansfield, *A History of the Middle East* (London: Penguin Books, 1992), 306.

8. Malu Halasa, "Funny Precarious," *The Guardian, Weekend,* July 27, 2002.

Up Close:
Intimate Still-lifes

1. Gilbert Hage, *Ici & maintenant* ("Here and Now"), 2003. www.gilberthage.com.

Bibliography

Abdelwahab, Bouhdiba. *Sexuality in Islam*. Translated by Alan Sheridan. London/Beirut: Saqi Books, 2004.

Abdulhamid, Ammar. *Menstruation*. London/Beirut: Saqi Books, 2001.

Aoude, Ibrahim G. "The Political Economy of Syria under Asad–Book Reviews." Review of *The Political Economy of Syria under Asad*, by Volker Perthes. *Arab Studies Quarterly*, fall 1997.

Baker, Patricia L. *Islamic Textiles*. London: British Museum Press, 1995.

Barakat, Hoda. *The Tiller of Waters*. Translated by Marilyn Booth. Cairo: American University of Cairo Press, 2001.

Biedermann, Ferry. "Damascus Declaration for Democratic and National Change." *Financial Times*, October 18, 2005.

Boustany, Noura. "A Modernizer Challenges Syria's Old Order." *Washington Post*, July 30, 2004, sec. A.

Bressler, Karen W., Karoline Newman, and Gillian Proctor. *A Century of Lingerie: Revealing the Secrets and Allure of 20th Century Lingerie*. Edison, NJ: Chartwell Books, 1997.

Burns, Ross. *Monuments of Syria: A Historical Guide*. London: I. B. Tauris, 2000.

Cunnington, C. Willett and Phillis. *The History of Underclothes*. Mineola, NY: Dover Publications, 1992.

Damandan, Parisa. *Portrait Photographs from Isfahan: Faces in Transition 1920–1950*. London/Beirut: Saqi Books/Prince Claus Fund Library, 2004.

Fadel, Marie and Rafik Schami. *Damascus: Taste of a City*. London: Haus Publishing, 2005.

Fattah, Hassan M. "The New Ramadan: It's Beginning to Look a Lot Like . . ." *International Herald Tribune*, October 12, 2005.

Greijer, Agnes. *A History of Textile Art*. Stockholm: Pasold Research Fund with Sothey Parke Bernet, 1979.

Hack, Susan. "Arabian Nighties." *Salon*, October 8, 1999 www.salon.com.

Halliday, Fred. "The *Millet* of Manchester: Arab Merchants and the Cotton Trade." In *Nation and Religion in the Middle East*. London/Beirut: Saqi Books, 2000.

Harris, Jennifer. *Textiles, 5,000 Years: An International History and Illustrated Survey*. New York: Henry Abrams, 1993.

Hitti, Philip K. *Capital Cities of the Arab World*. Minneapolis, MN: University of Minnesota Press, 1973.

Irwin, Robert. "Did This Man Inspire Bin Laden?" *The Guardian*, November 1, 2001.

Jeffreys, Andrew, ed. *Emerging Syria 2005*. London: Oxford Business Group.

Kallas, Faris N. "The 10th Baath Party Congress in Syria." Master's thesis, University of Westminster, 2005.

Keenan, Bridget and Tim Beddow. *Damascus: Hidden Treasures of the Old City*. London: Thames and Hudson, 2001.

Leverett, Flynt. *Inheriting Syria: Bashar's Trial by Fire*. Washington, D.C.: Brookings Institution Press, 2005.

Lindisfarne, Nancy. *Dancing in Damascus: Stories*. Albany, NY: State University of New York Press, 2000.

Marie, Elizabeth. "A Conversation with Hoda Barakat." In *Al-Jadid: A Review and Record of Arab Culture and Arts* 8, no. 39 (Spring 2002).

Mernissi, Fatimah. *Beyond the Veil*, rev. edition. Bloomington, IN: Indiana University Press, 1987.

Miller, Judith. "Syria." In *God Has Ninety-Nine Names: Reporting from a Militant Middle East*. New York: Simon and Schuster, 1996.

Nazar/Noorderlicht: Photographs from the Arab World. Curated by Wim Melis. Groningen, Netherlands: Stichting Aurora Borealis/Noorderlicht, 2004.

Quataert, Donald. *Ottoman Manufacturing in the Age of the Industrial Revolution*. Cambridge: Cambridge University Press, 1993.

Randall, Nicholas and Helena Diez. *Syria Revealed*. Madrid: Media Minds, 2004.

Salamandra, Christa. *A New Old Damascus: Authenticity and Distinction in Urban Syria*. Bloomington, IN: Indiana University Press, 2004.

Salhani, Claude. "The Syria Accountability Act Taking the Wrong Road to Damascus." *Policy Analysis*, March 18, 2004.

Schoeser, Mary. *World Textiles: A Concise History*. London: Thames and Hudson, 2003.

Seifan, Samir. "Towards Reform." *Syria Today*, September, 2005.

Seitz, Charmaine. "Here Comes the Son: Can Bashar Assad Bring Peace and Prosperity to Syria?" *In These Times,* January 22, 2001.

Smith, Lee. "The Way We Live Now: 2-13-05; ENCOUNTER; A Liberal in Damascus." *New York Times Magazine*, February 13, 2005, sec. 6.

Stack, Megan K. "Art of the Impolitic in Syria." *Los Angeles Times,* August 3, 2004, sec. A.

Touma, Issa. *The 8th International Photography Gathering: Aleppo 2004*. Aleppo: Le Pont Gallery, 2004.

Warnock Fernea, Elizabeth and B. Q. Bezirgan. *Middle Eastern Muslim Women Speak*. Austin, TX: University of Texas Press, 1978.

Weir, Shelagh. *The Bedouin: Aspects of Material Culture of the Bedouin of Jordan*. London: World of Islam, Festival Trust, 1976.

Editors and Contributors

Editors

Malu Halasa is an editor and journalist. She is coeditor of *Creating Spaces of Freedom: Culture in Defiance*, *Transit Beirut: New Writing + Images*, *Kaveh Golestan 1950–2003: Recording the Truth in Iran*, and *Transit Tehran: Young Iran and Its Inspirations*. Former managing editor of the Prince Claus Fund Library, she was also a founding editor of *Tank* magazine. She lives in London and writes for the British press.

Rana Salam has been running her own London-based design practice, www.ranasalam.com, over ten years. The studio specializes in Middle Eastern popular art and street culture, drawing on colorful imagery and using the latest design technology to create a unique vision of graphic design and art direction that has been commissioned by Paul Smith, Liberty's, Harvey Nichols, Villa Moda, Boutique 1, and in the Victoria & Albert Museum, among others. Her work has been exhibited in Lebanon, Dubai, and in the United Kingdom by Arts Council England and Institute of International Visual Arts. She also lectures and has organized workshops on the visual communication and the art of contemporary Middle East pop culture for the British Museum's Offscreen project, the 2007 International Design Forum in Dubai, and the British Council as part of Creative Lebanon.

Assistant Editor

Mitchell Albert is a London-based freelance book and magazine editor. He is the editor of *PEN International* and an associate editor of *Steppe*, a magazine focusing on Central Asia. He has worked for a number of publishing houses, magazines, and nonprofit organizations in the United Kingdom, United States, and elsewhere. His current projects include editing the war diaries of a private intelligence operative in Iraq and Afghanistan. In addition to having written or co-authored screenplays and teleplays, he is also an instructor of the Alexander Technique.

Contributors

Eugenia Dolberg is a photographer. She has worked throughout Asia and the Middle East for the past decade. She began with participatory photography in Cambodia and went on to found and direct Open Shutters, a program that trains women to share their experiences through photography and writing, in Syria and Iraq. She is based in Tehran and London.

Omar al-Moutem has been taking photographs for over twenty-five years. Initially taught at home by his elder brother, also a photographer, he first worked as a passport photographer and later as a product photographer before setting up his own studio. He photographs for Syrian lingerie manufacturers in Damascus.

Gilbert Hage is an artist and photographer who has exhibited in Lebanon, Brazil, and Germany. He participated in *Present Absence: Contemporary Art from Lebanon* at Galerie Tanit, Munich, and was invited by the Heinrich Böll Foundation to exhibit at the House of World Cultures, Berlin, for the 2005 conference "Identity Versus Globalization?" Hage teaches at the University Saint-Esprit de Kaslik, and the Académie Libanaise des Beaux-Arts (Alba), University of Balamand.

Iman Ibrahim has been writing poetry for over twelve years. Her poems have appeared in Syrian newspapers and magazines. Her first book of verse, *The Window of Zebra*, was published by the Syrian Ministry of Culture. After she studied photography at Le Pont Gallery in Aleppo, her work has consisted primarily of self-portraiture. She has exhibited in Syria, Lebanon, and Europe.

Noura Kevorkian is a filmmaker, painter, photographer, and composer. Her first documentary, *Veils Uncovered,* won numerous awards, including the National Film Board Award for Outstanding Canadian Documentary at the 2002 Reelworld Film Festival, Toronto, and the Golden Sheaf Award for Outstanding Documentary at the 2002 Yorkton Short Film & Video Festival. It was also nominated for Best Documentary at the 2002 New York International Independent Film and Video Festival.

Reine Mahfouz, a photographer, documented Palestinian camps in Lebanon for UNESCO. Her many ongoing photography projects include "Beirut Veiled," documenting the city's continuing evolution through its construction awnings, and the portraiture project "Nomadic Studio" featured at Home Works III, Ashkal Alwan's art symposium. In 2005, Mahfouz documented graffiti, including graffiti by the militias operating in Lebanon, in photographs published by the culture magazine *Zawaya*, in Beirut.

Issa Touma has been a photographer for the past fifteen years. In 1992, he established the first photography gallery in Syria, the Black and White Gallery. After its closure in 1995, he opened Le Pont, the only gallery dedicated to contemporary photography in Aleppo. In 1997, he founded the Aleppo biennial International Photography Gathering, an event that grew from six hundred visitors in its first year to more than seven thousand in 2004. In 1999, he organized the biennial International Women's Art Festival. In 2005, he was included in the Noorderlicht group exhibition, *Nazar: Photographs from the Arab World*. For the past decade he has been documenting Sufi festivals in Syria.

Index

A
Abdulhamid, Ammar, 7, 57–67
Aboud, Samir, 25
al-Adnan, 14, 16
al-Ali, Sumaya, 12, 25
Al Batul, 15
Aleppo, 9, 28, 30, 37, 42, 49, 55, 63
al-Moutem, Omar, 16, 19, 20, 26, 30, 67, 127
al-Tharwa, 57
Amasel Fashion, 14, 17, 20, 23, 127, 140–43
Angel Lady, 26, 127
Assad, Bashar, 28, 37
Assad, Hafez, 11, 30, 37
Asseel, 89
Avakian, Palig, 28
Azem, Zina, 81

B
Baqdounis, Hamas, 13
Basha, Layla, 116
Beirut, 9, 11, 44
belly dancing, 48, 49, 51
bikinis, edible, 16, 22
bras
   alternatives to, 29
   history of, 29
   sizing for, 26

C
Chantel Lingerie, 18, 19–20, 21
chocolate, 19

D
Daeih, Yousef, 25–26
Damascus
   changes in, 9, 11, 19, 66
   manufacturing in, 11, 63
   prostitution in, 28

DarEmar, 57
Din Belal, Abu Ali Salah al, 29
divorce, 52, 60
Dolberg, Eugenie, 69
Doukmak, Mouhammed, 17, 20, 23, 26

E
Elias, Lina Makss, 108

F
Farah, 20
Farzat, Ali, 67
Fathi family, 36, 39–40, 41, 44, 45, 48, 51–53, 54
Flora, 89, 92

H
Habloul, Dareen, 25
Hadiths, 62
Hage, Gilbert, 9, 69
Haliby, Muhammad Emad, 14, 16
Hanin, 89
Hitti, Philip K., 9
homosexuality, 65
humor, 64–65

I
Ibrahim, Iman, 9, 155–65
Islam
   divorce and, 52
   fundamentalist, 57–58
   polygamy and, 44–45

K
Kabbani, Ghalia, 108
Karizma, 81
Kasar, Alaa, 20
Kevorkian, Noura, 8, 35

L

Lahham, Lena, 125
Lebanon, commercial ties with, 89
Le Pont, 30
lingerie, Syrian
   comfort and, 63
   fashion cycle for, 7
   history of, 28–30
   influences on, 7–8
   Muslim vs. Christian attitudes toward, 103
   product photography from manufacturers of, 8, 16, 17, 30, 126–53
   reasons for buying, 41, 44
   still-lifes of, 68–71, 73–80, 82–88, 90–91, 93–102, 104–7, 109–15, 117–24
   symbolism of, 8
   variety of, 7, 13–14, 32, 60
   weddings and, 13, 48, 61–62
   women's attitudes toward, 69, 72, 81, 89, 92, 103, 108, 116, 125
Lingerie Baqdounis, 13–14, 33
Lingerie Lour, 12, 25–26

M

Mahfouz, Reine, 8, 11
marriage, 60
men's tailoring, 63
menstruation, 57–58
Menstruation (novel), 57, 59
modeling, 8, 30, 126–53
Mousali family, 15
Muhammad, sayings attributed to, 62
Murad, Khalil, 19, 21

N

Nabulsi, Firas, 26, 28

O

October War, 28
Odabashi, Muhammad Zuhair, 19
Omari family, 15

P

polygamy, 44–45
pornography, 65–66
prostitution, 28

Q

Qutb, Sayyid, 58

R

Ramadan, Hassan, 69
Raqqah, 49

S

sex tourism, 66–67
sexuality
   humor and, 64–65
   issues with, 59, 62, 65, 67
Shaheen, Mohammed, 57
Shenineh, 29–30
Shenineh, Yasser, 30
Shirley, 20
Sioufi, Shadi, 25
Souk al-Hamidiyeh, 12, 13, 15, 35, 54, 61
Syria
   commercial ties between Lebanon and, 89
   as country in transformation, 9, 60–61, 89
   humor in, 64–65
   manufacturing and, 63
   marriage and divorce in, 60
   polygamy in, 44
   sexuality in, 59
   textile industry of, 11, 13, 29–30
   Western culture and, 58–59
   women's status in, 8–9, 44, 59–60, 116, 125

T
Tabbakh, Nihad, 28-29
Temizian, Taleen, 89
Touma, Issa, 30, 35

U
Umayyad Mosque, 11

V
Veils Uncovered (film), 8, 35, 53
Victoria's Secret, 25, 63, 92

W
Wasef, Mona, 57
weddings, 13, 48, 61-62
women
    attitudes of, toward lingerie, 69, 72, 81, 89, 92, 103, 108, 116, 125
    divorced, 52, 60
    menstruating, 57-58
    Muslim vs. Christian, 103
    status of, 8-9, 44, 59-60, 116, 125

Acknowledgments

This book would not have been possible without a generous grant from the Prince Claus Fund, The Hague. The editors thank Els van der Plas, director of the Prince Claus Fund; Peter Stepan, the managing editor of the Prince Claus Fund Library; and the library's advisory editorial board: Emile Fallaux, Ian Buruma, Ellen Ombre, and Okwui Enwezor. In Damascus, Fares Kallas provided invaluable support. In London, Emily Campbell at the British Council offered an early research opportunity by sending the editors to Creative Lebanon (2003). Lena Lahham was generous with her assistance, as was Nupu Press and Walter Keller. Iman Ibrahim's poetry was translated by Sarah al-Hamad, assisted by Hassan Ramadan and Zaid Awdat. The editors also thank the lingerie manufacturers of Syria, Bridget Watson Payne at Chronicle Books, and Assem Salam.